Soil Quality Standards for Trace Elements

Derivation, Implementation, and Interpretation

Other Titles from the Society of Environmental Toxicology and Chemistry (SETAC)

Semi-Field Methods for the Environmental Risk Assessment of Pesticides in Soils
Schaeffer, van den Brink, Heimbach, Hoy, de Jong, Römbke, Roß-Nickoll, Sousa
2010

Ecotoxicology of Amphibians and Reptiles
Sparling, Linder, Bishop, Krest, editors
2010

Ecological Assessment of Selenium in the Aquatic Environment
Chapman, Adams, Brooks, Delos, Luoma, Maher, Ohlendorf, Presser, Shaw, editors
2010

Application of Uncertainty Analysis to Ecological Risks of Pesticides
Warren-Hicks and Hart, editors
2010

Risk Assessment Tools Software and User's Guide
Mayer, Ellersieck, Asfaw
2009

Derivation and Use of Environmental Quality and Human Health Standards for Chemical Substances in Water and Soil
Crane, Matthiessen, Maycock, Merrington, Whitehouse, editors
2009

Linking Aquatic Exposure and Effects: Risk Assessment of Pesticides
Brock, Alix, Brown, Capri, Gottesbüren, Heimbach, Lythgo, Schulz, Streloke, editors
2009

Aquatic Macrophyte Risk Assessment for Pesticides
Maltby, Arnold, Arts, Davies, Heimbach, Pickl, Poulsen
2009

For information about SETAC publications, including SETAC's international journals, Environmental Toxicology and Chemistry and Integrated Environmental Assessment and Management, contact the SETAC office nearest you:

SETAC
1010 North 12th Avenue
Pensacola, FL 32501-3367 USA
T 850 469 1500 F 850 469 9778
E setac@setac.org

SETAC Office
Avenue de la Toison d'Or 67
B-1060 Brussells, Belguim
T 32 2 772 72 81 F 32 2 770 53 86
E setac@setaceu.org

www.setac.org
Environmental Quality Through Science®

Soil Quality Standards for Trace Elements

Derivation, Implementation, and Interpretation

Edited by
Graham Merrington and Ilse Schoeters

Coordinating Editor of SETAC Books
Joseph W. Gorsuch
Copper Development Association, Inc.
New York, NY, USA

CRC Press is an imprint of the
Taylor & Francis Group, an **informa** business

Information contained herein does not necessarily reflect the policy or views of the Society of Environmental Toxicology and Chemistry (SETAC). Mention of commercial or noncommercial products and services does not imply endorsement or affiliation by the author or SETAC.

CRC Press
Taylor & Francis Group
6000 Broken Sound Parkway NW, Suite 300
Boca Raton, FL 33487-2742

First issued in paperback 2019

ISBN-13: 978-1-4398-3023-9 (hbk)
ISBN-13: 978-0-367-38346-6 (pbk)

Visit the Taylor & Francis Web site at
http://www.taylorandfrancis.com

and the CRC Press Web site at
http://www.crcpress.com

SETAC Publications

Books published by the Society of Environmental Toxicology and Chemistry (SETAC) provide in-depth reviews and critical appraisals on scientific subjects relevant to understanding the impacts of chemicals and technology on the environment. The books explore topics reviewed and recommended by the Publications Advisory Council and approved by the SETAC North America, Latin America, or Asia/Pacific Board of Directors; the SETAC Europe Council; or the SETAC World Council for their importance, timeliness, and contribution to multidisciplinary approaches to solving environmental problems. The diversity and breadth of subjects covered in the series reflect the wide range of disciplines encompassed by environmental toxicology, environmental chemistry, hazard and risk assessment, and life-cycle assessment. SETAC books attempt to present the reader with authoritative coverage of the literature, as well as paradigms, methodologies, and controversies; research needs; and new developments specific to the featured topics. The books are generally peer reviewed for SETAC by acknowledged experts.

SETAC publications, which include Technical Issue Papers (TIPs), workshop summaries, newsletter (*SETAC Globe*), and journals (*Environmental Toxicology and Chemistry* and *Integrated Environmental Assessment and Management*), are useful to environmental scientists in research, research management, chemical manufacturing and regulation, risk assessment, and education, as well as to students considering or preparing for careers in these areas. The publications provide information for keeping abreast of recent developments in familiar subject areas and for rapid introduction to principles and approaches in new subject areas.

SETAC recognizes and thanks the past coordinating editors of SETAC books:

A.S. Green, International Zinc Association, Durham, North Carolina, USA
C.G. Ingersoll, Columbia Environmental Research Center, US Geological Survey, Columbia, Missouri, USA
T.W. La Point, Institute of Applied Sciences, University of North Texas, Denton, Texas, USA
B.T. Walton, US Environmental Protection Agency, Research Triangle Park, North Carolina, USA
C.H. Ward, Department of Environmental Sciences and Engineering, Rice University, Houston, Texas, USA

Contents

List of Figures

List of Tables

Acknowledgments

The Technical Workshop on deriving, implementing, and interpreting soil quality standards for trace elements and publication of the workshop report were made possible through the financial support of the International Zinc Association (IZA), European Copper Institute (ECI), Eurometaux, Nickel Producers Environmental Research Association (NiPERA), International Lead Zinc Research Association (ILZRO), International Council on Mining and Metals (ICMM), Cobalt Development Institute (CDI), Rio Tinto, Anglo American, International Molybdenum Association (IMOA), Vale INCO, Teck Cominco Ltd., Environment Agency of England and Wales, Department of Environment, Climate Change and Water NSW, Metals in the Human Environment Strategic Network (MITHE-SN), and CSIRO Land and Water.

The content of this publication does not necessarily reflect the position or the policy of any of these organizations.

The workshop organizers gratefully acknowledge the participants for their enthusiastic commitment to the workshop's objectives.

A special word of thanks goes to Sandra Tyrrell for taking care of all local arrangements and logistical matters.

We also acknowledge Steve McGrath, who provided the peer review for this report, and Joe Gorsuch for his support in editing the chapters of this report.

Lastly, we acknowledge the guidance and unstinting support of Mimi Meredith in the production of this book.

About the Editors

After receiving a BSc in Environmental Science and PhD in the environmental behavior of metals at historic mine sites from Queen Mary College London, **Graham Merrington** took up a postdoctoral research position at the Department of Soil Science, University of Reading. His research activities were directed toward assessing the fate and behavior of metals in contaminated soils and wastes, specifically in relation to the influence of organic carbon. From Reading, Graham moved to Bournemouth in 1994, where he took up a Lectureship in Environmental Chemistry and teamed up with colleagues to look at the transfer of metals through terrestrial food chains, specifically the soil-plant-insect linkage. In 1998 Graham took up a position at Adelaide University as a Lecturer in Soil Chemistry, where he continued his work on metal behavior with the help of colleagues at CSIRO. Dr. Merrington has more than 50 scientific publications focusing on the behavior and fate of metals in terrestrial and aquatic systems.

Graham returned from Australia in 2002 to join the Environment Agency of England and Wales where he led an R&D Program focused on Environmental Quality Standards in soils, waters, and sediments. Key projects included the incorporation of Biotic Ligand Models into compliance assessment for regulators, the implementation and use of ecologically based soil standards for metals, and the derivation of soil quality indicators to assess sustainable land management. He also represented the United Kingdom at Expert Groups for the Water Framework Directive and was a regular attendee as an expert at TCNES for metals related issues. Graham is now a director at wca environment, an independent research and consultancy company established in 2005 by experienced chemical risk assessors, providing assistance to industry and governmental agencies in the field of environmental toxicology and risk assessment.

Ilse Schoeters is principal advisor regulatory affairs for Europe, the Middle East, and Africa, at Rio Tinto Minerals, Belgium. She was trained as an agricultural engineer with specialization in soil science. She started her career as a researcher at the Catholic University of Leuven, where she studied the overfertilization of soils in Flanders and was involved in the European risk assessment of cadmium. From there she moved on to Eurometaux, the European nonferrous metal association, where she was responsible for the scientific aspects of the environmental legislations on metals. She also worked at the European Copper Institute on environmental and health regulations. Major projects she worked on included development of a model to predict the toxicity of copper in soils, where she served as the research coordinator, and the European risk assessment for copper, where she was responsible for all soil-related aspects of the assessment. Current work at Rio Tinto includes assessing toxicity of borates in soils.

Workshop Participants*

Graham Merrington wca environment Ltd, Oxfordshire, United Kingdom
Ilse Schoeters Rio Tinto, Gent, Belguim
Monica J.B. Amorim Centre for Environmental and Marine Studies (CESAM), University of Aveiro, Aveiro, Portugal
Nick Basta Ohio State University, Columbus, Ohio, USA
Andreas Bieber Ministry for the Environment, Nature Conservation and Nuclear Safety, Bonn, Germany
Craig Boreiko International Lead Zinc Research Organization (ILZRO), Durham, North Carolina, USA
Teresa Bowers Gradient Corporation, Cambridge, Massachusetts, USA
Lucia Buvé Umicore EHS, Brussels, Belgium
John Chapman Department of Environment & Climate Change & Water, Lidcombe, New South Wales, Australia
Victor Dries Public Waste Agency of Flanders (OVAM), Mechelen, Belgium
Anne Fairbrother Parametrix, Corvallis, Oregon, USA
Peter Glazebrook Rio Tinto
Wang Guoqing Center for Assessment and Remediation of Contaminated Sites, Nanjing Institute of Environmental Science (NIES), Ministry of Environmental Protection of the People's Republic of China, Beijing, China
Beverley Hale University of Guelph, Guelph, Ontario, Canada
Diane Heemsbergen Commonwealth Scientific and Industrial Research Organisation (CSIRO), Glen Osmond, South Australia, Australia
William Hendershot McGill University, Montreal, Quebec, Canada
Seung-Woo Jeong Kunsan National University, Kunsan, Korea
Andrew Langley Sunshine Coast Council Public Health Unit, Maroochydore, Queensland, Australia
Roman Lanno Ohio State University, Columbus, Ohio, USA
Betty Locey Arcadis, Novi, Michigan, USA
Steve Lofts Center for Ecology and Hydrology (CEH), Lancaster Environment Centre, Lancaster, United Kingdom
Yibing Ma Chinese Academy of Agricultural Sciences (CAAS),
Mike J. McLaughlin Commonwealth Scientific and Research Organization (CSIRO) and University of Adelaide, Adelaide, South Australia, Australia

* Affiliations were current at the time of the workshop.

Co Molenaar	Ministry of Housing, Spatial Planning and the Environment (VROM), Den Haag, Netherlands
Michael Moore	Australian National Research Centre for Environmental Toxicology (EnTOX), Queensland University, Brisbane, Queensland, Australia
Marylène Moutier	SPAQUE, Liege, Belguim
Chris Oates	Anglo American plc, London, United Kingdom
Graeme Paton	University of Aberdeen, Aberdeen, United Kingdom
Jussi Reinikainen	Finnish Environment Institute, Helsinki, Finland
Leonard Ritter	University of Guelph, Guelph, Ontario, Canada
Chris Schlekat	Nickel Producers Environmental Research Association (NiPERA), Durham, North Carolina, USA
Erik Smolders	Katholieke Universiteit Leuven, Faculty of Applied Biosciences and Engineering, Heverlee, Belgium
Jaana Sorvari	Finnish Environment Institute, Helsinki, Finland
Gladys Stephenson	Stantec Consulting, Guelph, Ontario, Canada
Shu Tao	Laboratory for Earth Surface Processes, Peking University, Beijing, China
Michael St. J. Warne	Commonwealth Scientific and Research Organization (CSIRO), Glen Osmond, South Australia, Australia
Randy Wentsel	U.S. EPA Office of Research and Revelopment, Washington, D. C., USA

1 Workshop on Deriving, Implementing, and Interpreting Soil Quality Standards for Trace Elements

Graham Merrington and Ilse Schoeters

1.1 INTRODUCTION TO THE WORKSHOP

This book is the result of discussions that took place at a Society of Environmental Toxicology and Chemistry (SETAC)-sponsored Technical Workshop on deriving, implementing, and interpreting soil quality standards (SQSs) for trace elements (TEs). The workshop took place in July 2008, in Sydney, Australia, and built on the outputs of a number of other previous SETAC workshops, specifically, the recent workshop on the derivation and use of Environmental Quality Standards in soils and water (Crane et al. 2009) and the work by Carlon (2007).

The purpose of the workshop was to facilitate a focused discussion on the science and methodologies underpinning the derivation of SQSs for TEs. Specifically, it provided a common forum for environmental regulators, scientists, and environmental managers to share their views and understand how concepts such as (bio) availability and exposure modeling should be used in setting and using soil standards for the protection of the environment and human health. While complete harmonization of standards may be some way off, key paradigms and concepts in delivering implementable and interpretable metrics by which to routinely assess potential TEs' risks in soils should be an achievable goal (Carlon 2007). Achieving this goal would deliver gains in efficiency, consistency, and practicality for businesses and regulators.

Thirty-eight experts from 11 countries from Europe, Asia, and North America representing a multidisciplinary group of government policy makers and regulators, academics, industry representatives, and consulting firms met for 3.5 days of discussions on the science underpinning the best practice for deriving and implementing quality standards for TEs in soils.

The first half day of the workshop was spent in a plenary session, designed to review key issues, cross-cutting themes, and objectives of the workshop. This included scene-setting presentations on

1) the range of environmental and human health TE standards applied worldwide;
2) the technical, legal, and regulatory frameworks that underpin the derivation and use of standards worldwide, but with examples from European Member States, Canadian, US, Australian, and Chinese regulations; and
3) key issues from current international research on environmental and human health standards for TEs.

Participants were assigned to one of three work groups in order to tackle some broad themes required to achieve the workshop aims. The work groups and their aims were as follows:

1) Derivation of ecological-based terrestrial standards for TEs:
   - evaluating data quality and methods of standard derivation,
   - considering how the bioavailability of TEs for soil organisms should be used when deriving or using standards, and
   - reviewing how uncertainty should be taken into account when setting standards at regional, national, and site-specific scale.
2) Derivation of human health-based terrestrial standards for TEs:
   - evaluating data quality and methods for human health standards for TEs,
   - considering how the bioavailability of TEs for humans and vertebrates should be used when deriving or using standards,
   - reviewing how uncertainty should be taken into account when setting standards at regional, national, and site-specific scale, and
   - reviewing how standards can be validated or verified.
3) Implementation and use of terrestrial standards for TEs:
   - reviewing methods to assess compliance with SQSs,
   - evaluating how the requirements for deriving soil standards should be interpreted when considering site-specific assessments and/or site-specific standards,
   - considering how the bioavailability of TEs can be measured or estimated when assessing risks in the field, and
   - considering how we can move toward an international, common "toolbox" for risk evaluation of soil contamination.

Participants were asked to aim for consensus on scientific issues relevant to the derivation and use of SQSs for TEs. Furthermore, it was recommended by the organizing committee that the participants question the assumptions used in the SQS derivation process by asking questions that would clarify and define the boundary between policy and science, and in doing so, they should assume good faith on the part of others.

The results of the workshop are contained in this monograph. Chapter 2 outlines the underpinning science when deriving SQSs for TEs aiming to protect soil fauna and flora. Chapter 3 undertakes the same but for SQSs aiming to protect human health. Chapter 4 covers a topic area that is often overlooked by the scientific community, and that is the application and practical use of SQSs for TEs by environmental managers and regulators in assessing potential risks. Finally, in Chapter 5, we draw together the overall conclusions of the workshop and provide recommendations on the development and use of SQSs for TEs. We also identify future research that would help to underpin the science of environmental and human health standards.

It should be noted that the discussions were focused on soil fauna and flora and human health as protection goals. Groundwater was not included because the workshop participants agreed this required expertise was not available among the participants.

1.2 DERIVING, IMPLEMENTING, AND INTERPRETING SQS FOR TEs

Quality standards are widely used to protect various compartments of the environment (e.g., water, soil, and sediment) and human health from chemicals released by human activity. Generally, quality standards relate to doses or concentrations in the environment for specific chemicals, below which unacceptable effects are not expected to occur.

In some jurisdictions there may be more than 30 years of experience in the setting and use of quality standards for the aquatic compartment; yet the development of quality standards for chemicals in the terrestrial compartment is a relatively new regulatory activity applied in a limited number of countries. In the European Union (EU), for example, only 9 Member States have specific legislation on soil protection, while the others rely on a number of indirect policy provisions to safeguard soils (Carlon 2007).

Quality standards for chemical substances in groundwater and soil form the basis for many soil quality decisions, including emission reduction measures during the admission of chemicals on the market, land management decisions, risk management, and soil remediation.

The protection goals of SQSs, derivation methods, and frameworks within which they are used differ between countries and regions. This diversity reflects genuine technical differences that must be taken into account in the development of standards for different soil types or for different receptors (e.g., humans, livestock, or soil flora and fauna). However, much standard setting has been developed in a piecemeal fashion with limited consistency in the levels of protection sought (even within the same country) or the scientific data and methods used to derive or interpret them. Furthermore, the methods used to monitor compliance differ among countries or regions. These differences can lead to the implementation of substantially different values and of different risk assessment results using the same empirical data and the same protection goals, which must mean that their application is either over- or underprecautionary in at least some situations. The ramifications of inappropriate

SQSs may be considerable. For example, along with environmental implications there are also economic and social effects on business and property development.

There is a significant national and international legislative drive to establish SQSs for TEs for the protection of various compartments of the environment and human health. The need for harmonization of the methods to derive SQSs and assess risks is promoted by the EU in the proposed Soil Framework Directive. Other countries (e.g., China, Australia, and New Zealand) are currently in the process of refining and revising SQS for contaminants.

The term "trace element" has several meanings depending on the context in which it is used. For analytical chemists it can be an element detected in a sample at concentrations of less than 100 mg kg^{-1} (IUPAC 2006), and for geochemists and some soil scientists it can be taken as an element that has a concentration in the soil or rock of less than 1000 mg kg^{-1}. In this book, TEs include micronutrients and non-essential elements, e.g., Fe, Mn, Zn, B, Cu, Mo, As, Hg, Be, Ni, Sb, Se, Cd, Ag, Cr, Co, Tl, Cu, Zn, and Pb.

Historically, the substances for which SQSs have been set nearly always include TEs. Yet TEs have a unique set of characteristics that can make the implementation and use of SQSs especially onerous. These characteristics include variable "naturally" occurring background concentrations, differences in (bio)availability and toxicity between soils with different properties, and the essentiality of some of these elements.

Catalyzed by regulatory programs such as the EU Existing Substances Regulation, the science to assess hazards and risks of TEs in soils has advanced significantly over the past 10 years. Recent developments in the understanding of TE fate and behavior in soils has led to a recognition that the use of traditional strong acid digestion methods are a poor basis for SQSs and thus to assess ecological and human health risks. A SETAC Pellston workshop (Fairbrother et al. 2002) evaluated the state of the science on hazard assessment of sparingly soluble metal compounds in soils and made recommendations on how to improve the toxicity test methods and how to use this information for hazard and risk assessments. Further building on this theme was the SETAC Pellston workshop on hazard identification approach for metals and inorganic metal substances, which directly addressed the assessment of hazards in terrestrial system (Adams and Chapman 2003).

More recently, and of direct relevance to this meeting, was the SETAC-sponsored Pellston Workshop (Crane et al. 2009), which discussed the broad issues of environmental quality standard setting including the terrestrial compartment. The workshop provided generic guidance and reasoning on appropriate ways forward to improve the implementability of environmental quality standards. Scientific, social, political, and economic considerations were discussed to deliver standards that meet the range of protection goals in different environmental compartments. Finally, Carlon (2007) made a comprehensive review of the differences in deriving SQS between the EU Member States.

This workshop has built on the outputs of these previous meetings, aiming specifically to discuss in depth the scientific aspects related to the setting and implementing of SQSs for TEs and to provide recommendations for the harmonization of the scientific basis of SQSs.

1.3 AIMS AND OBJECTIVES OF THE MEETING

The purpose of this workshop was to establish an understanding of the status of science related to ecological and human health protection from TEs in soils. Each work group would discuss the underpinning approaches and considerations of international jurisdictions on the derivation and use of soil standards for TEs. This would lead to the identification and promotion of "best practice" in accounting for (bio) availability and exposure modeling in standard setting for soils. Finally, the best practice was to be contextualized through an assessment of the future directions and developments in implementing TE standards for soils.

REFERENCES

Adams WJ, Chapman PM, editors. 2003. Assessing the hazard of metals and inorganic metal substances in aquatic and terrestrial systems. Pensacola (FL): SETAC Pr.

Carlon C. 2007. Derivation methods of soil screening values in Europe: a review and evaluation of national procedures towards harmonisation. Report EU 22805-EN. Ispra (Italy): European Commission, Joint Research Centre. 306 p.

Crane M, Matthiessen P, Stretton Maycock D, Merrington G, Whitehouse P, editors. 2009. Derivation and use of environmental quality and human health standards for chemical substances in water and soil. Pensacola (FL): SETAC Pr.

Fairbrother A, Glazebrook PW, Tarazona JV, van Straalen NM, editors. 2002. Test methods to determine hazards of sparingly soluble metal compounds in soils. Pensacola (FL): SETAC Pr.

IUPAC. 2006. Compendium of Chemical Terminology, 2nd ed. (the "Gold Book"). Compiled by A. D. McNaught and A. Wilkinson. Oxford (UK): Blackwell Scientific, (1997). XML online corrected version: http://goldbook.iupac.org (2006–) created by Nic M, Jirat J, Kosata B; updates compiled by A. Jenkins. ISBN 0-9678550-9-8. doi:10.1351/goldbook.

2 Derivation of Ecologically Based Soil Standards for Trace Elements

Mike J. McLaughlin, Steve Lofts, Michael St. J. Warne, Monica J.B. Amorim, Anne Fairbrother, Roman Lanno, William Hendershot, Chris E. Schlekat, Yibing Ma, and Graeme I. Paton

2.1 INTRODUCTION

There are many challenges in deriving ecologically based soil quality standards (SQSs) for trace elements (TEs) in soils due to the variable nature of soils across sites, regions, and continents and due to the variable nature of the ecological endpoints that need to be protected. Different jurisdictions use different terminology for SQS, e.g., soil screening values, ecological investigation levels, maximum permitted concentrations, trigger values, etc., but in this chapter we will use the term SQS and define it to mean a threshold concentration of TE in soil above which some action is required (e.g., further investigation, toxicity assessment, remediation, etc.) (Chapter 4). Generally, the preferred use of SQSs is to trigger further investigation, and risk management is usually only triggered following further investigation and derivation of site- or land use-specific SQSs against which the dose is compared (Carlon 2007).

2.2 SOIL FACTORS AFFECTING EFFECTIVE DOSE

2.2.1 Background Concentrations

The concentrations of TEs in soil are derived from both soil parent material (geological parent rock, loess, or sediment—geogenic sources) and TE inputs to soil from urban, industrial, and agricultural sources (anthropogenic sources). Concentrations of TEs derived from geogenic sources are generally regarded as "background concentrations." The concept of background concentrations and the definitions of these have already been thoroughly discussed by Reimann and Garrett (2005). It is important to note that due to long-range atmospheric transport of anthropogenic sources of TEs, "pristine" or "natural" background concentrations of most TEs (i.e., in soil totally unimpacted by human activity) rarely exist now. Hence "ambient background concentration" (ABC)

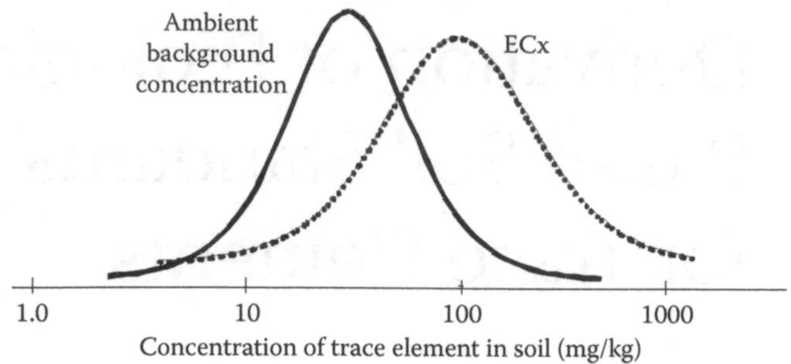

FIGURE 2.1 Schematic representation of overlap of ambient background concentrations of TEs in soil and effect concentrations (ECx) values.

is the term more commonly used to describe the concentration of TE in a soil that is distant from urban and industrial sources of TE and has not had large additions of TE in fertilizers, wastes, manures, or other soil amendments (Zhao et al. 2007).

ABCs of TEs in soil provide a dilemma for many regulatory agencies trying to protect soils from TE pollution (Chapter 4). Quite often, the distribution of ecological threshold effect values (ECx) overlap significantly with the natural variability in ABCs across a regulatory jurisdiction, region, or continent (Figure 2.1).

Because of the large range of ABCs of TEs in soil and because of the desire to set SQSs on the basis of species sensitivity distributions (SSDs, discussed below), this caused many early proposed SQSs (using total TE concentrations in soil) to be below ABCs (van de Meent et al. 1990). For a range of TEs, this overlap is considerable and especially problematic for As, Cd, Cr, Hg, and Ni if we compare typical background concentrations and the range of SQSs currently used in Europe (Figure 2.2).

SQS values triggering investigation rather than cleanup (at the lower end of the boxes in Figure 2.2) are often the ones closest to geogenic background values. SQS values triggering remediation (predicting significant ecological or human risk) are usually above both geogenic background concentrations and ABCs.

To overcome this problem, the "added risk approach" was developed (Struijs et al. 1997), where the total TE concentration in soil was divided into "background" (C_b) and "anthropogenic" (C_a) concentrations. If geogenic background concentrations of TEs were assumed to have no effects on biota because of negligible bioavailability and/or because the organisms have adapted to these concentrations, then SQSs can be developed by setting standards for added TE (C_a), to which the geogenic background concentration is added (C_b, equivalent to the ABC) to produce a threshold total TE concentration ($C_b + C_a$) (Figure 2.3a). The permissible TE addition (C_a) is defined using SSDs or other approaches that consider only added TE concentrations in toxicity tests. However, it was recognized that this was a simplistic view of TE availability in soil and that both the geogenic and anthropogenic fractions of TE in soil could have bioavailable (active) and nonbioavailable (inactive) fractions (Figure 2.3b).

The added risk approach has subsequently been used to determine SQS for a wide range of TEs (Crommentuijn et al. 2000a, 2000b; Sijm et al. 2001; Heemsbergen et al. 2008; Vlaams Reglement Bodemsanering [VLAREBO] 2008; Heemsbergen et al. 2009a).

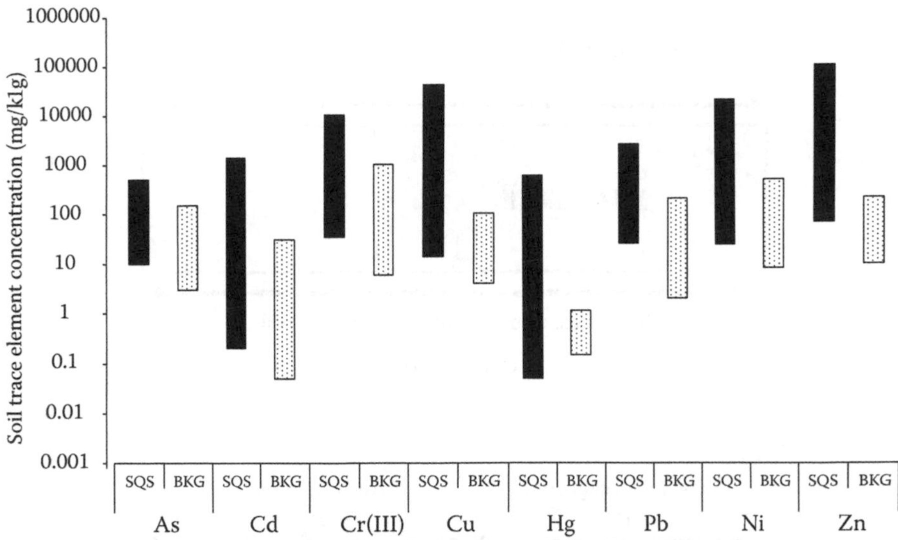

FIGURE 2.2 Range of ambient background (BKG) concentrations of TEs in soil (from McLaughlin, 2002) compared to range of SQSs for the same elements across Europe.

Limitations of the added risk approach include the difficulty in distinguishing between added and geogenic background TE in any given soil sample and also how to estimate or directly measure the fraction of the geogenic background TE that is bioavailable. These limitations and others have led some organizations to adopt the total risk approach, which bases the derivation of SQSs on total TE concentrations with suitable bioavailability and species sensitivity corrections. The added risk and total risk approaches are described in The Metals Environmental Risk Assessment Guidance (MERAG) document, which shows that both can incorporate many of the sophisticated approaches to estimate the bioavailable fraction (e.g., bioavailability normalization, leaching/aging correction, species sensitivity correction) that are discussed in later sections of this chapter (http://www.icmm.com/document/258).

Regardless of the approach, it is important to understand the distribution of ABCs in terms of implementing SQS values and determining their compliance. Choosing a single ABC is therefore difficult owing to the huge variation in ABCs regionally and across continents. At present, 3 different approaches have been suggested for the determination of ABCs for TEs in soil:

1) Regional or continental surveys of soils impacted minimally by human activity and choice of a concentration that represents a defined percentile of the distribution of soil concentrations measured (Zarcinas et al. 2003, 2004; International Standards Organization (ISO) 2005). Indeed, this is the technique suggested by the ISO (2005). "Outlier" points can be identified and then excluded using various formulae (ISO 2005; Zhao et al. 2007; European Chemicals Agency [ECHA] 2009) or by using box and whisker plots (Figure 2.4).

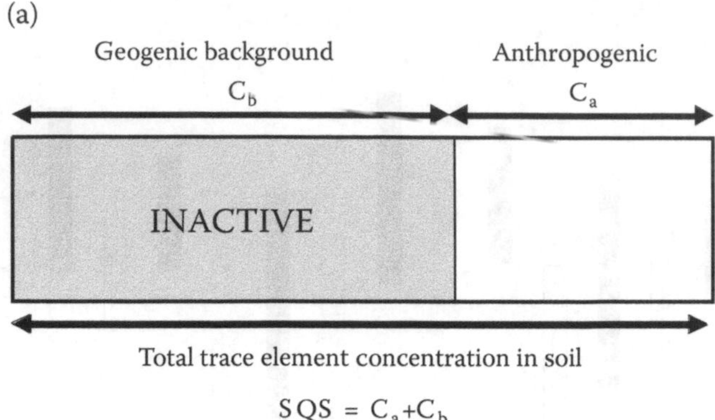

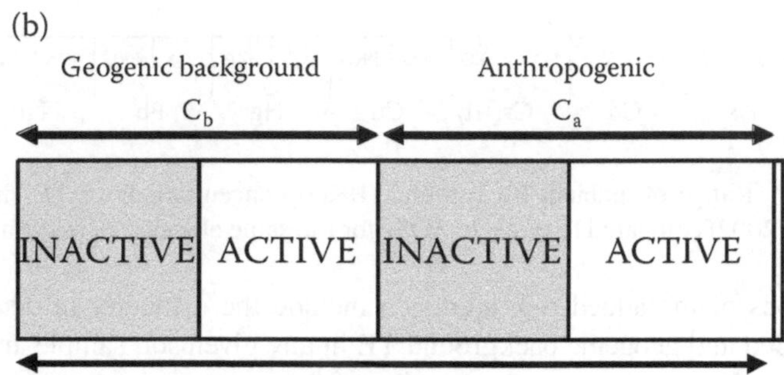

FIGURE 2.3 Schematic of the added risk approach where maximum permissible concentrations of contaminants in soils are determined assuming (a) that the geogenic background contaminant (C_b) is not bioavailable (inactive), or (b) that both the geogenic background and added TE have bioavailable (active) and non-bioavailable (inactive) components. (Adapted from Struijs et al. 1997.)

2) Probability function methods, where anthropogenic contamination of soils is suspected in the regional surveys. Here deviation from (log) normality at higher TE concentrations is assumed to be due to anthropogenic contamination. It is assumed the data have a lognormal distribution and the ABCs are selected by choosing a defined percentile of the forced distribution of values (Figure 2.5). Thus any data that are above the defined ABC value are assumed to have received extensive anthropogenic contamination.
3) Geochemical regression methods—to avoid the use of regionally or continentally defined ABCs, several authors have attempted to normalize TE concentrations to soil physical characteristics such as clay content (Lexmond and Edelman 1994; Zhao et al. 2007) or to concentrations of structural elements in soil such as Al, Fe, or Mn important in TE retention against leaching and biological removal (Hamon et al. 2004; Oorts et al. 2006; Zhao et al. 2007) (Table 2.1). The regression models developed by these

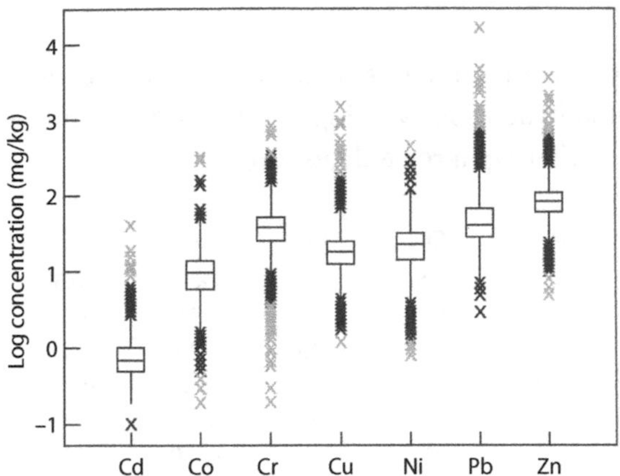

FIGURE 2.4 Box plots for the log transformed data of cadmium (Cd), cobalt (Co), chromium (Cr), copper (Cu), nickel (Ni), lead (Pb), and zinc (Zn). The rectangular blocks represent the 25th–75th percentiles, the horizontal lines within the boxes represent the median values, and the vertical lines outside the boxes represent the lower and upper whisker. Black and grey crosses are outliers and far outliers, respectively (reprinted from Environmental Pollution 148/1, Zhao et al. Estimates of ambient background concentrations of trace metals in soils for risk assessment, Copyright (2007), with permission from Elsevier).

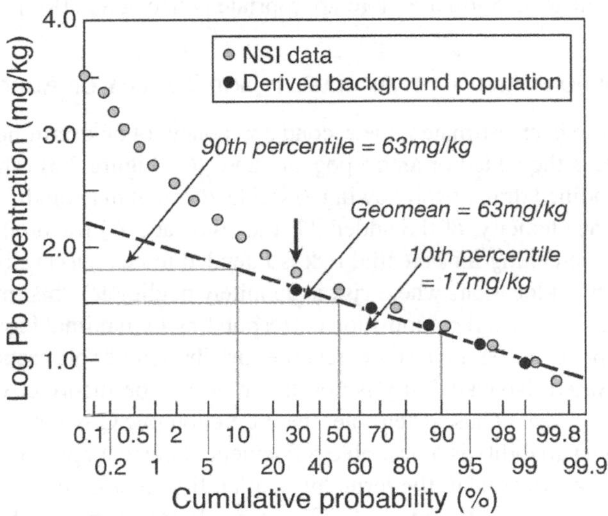

FIGURE 2.5 Probability graph for soil Pb concentrations in England and Wales, and extrapolated line shown assuming a log-normal distribution. NSI=National (UK) Soil Inventory data. The derived background population was assumed to be from "uncontaminated soils" assuming the inflexion of the NSI data curve occurred at 30% cumulative probability (reprinted from Environmental Pollution 148/1, Zhao et al., Estimates of ambient background concentrations of trace metals in soils for risk assessment, Copyright (2007), with permission from Elsevier).

TABLE 2.1
Upper expected (95th percentile) trace element concentrations in uncontaminated soils varying as a function of soil Fe content (measured after aqua regia digestion)[a]

Soil Fe (%)	As	Cr	Cu	Ni	Pb	Zn
0.1	<3	<15	<4	<5	<0.3	<9
0.5	<8	<50	<10	<15	<2	<25
1	<10	<80	<15	<25	<4	<35
5	<30	<275	<45	<75	<20	<85
10	<45	<465	<70	<120	<40	<130
25	<75	<925	<120	<230	<100	<225

[a] Concentrations are in mg kg^{-1}. Content measured after aqua regia digestion. Data from Hamon et al. (2004).

methods become less robust as the geographic area (and hence variation in geochemical parent rock) increases; i.e., these relationships work best on smaller geographic areas. Choice of percentile used is also important—in the approach 95th percentiles are not conservative, so perhaps lower percentile data (e.g., 50th) are more appropriate (Zhao et al. 2007).

2.2.2 How Soils Affect the Availability and Toxicity of Added TEs

After ABCs have been estimated, the second component of determining the effective dose of the TE is the added or anthropogenic dose (C_a, Figure 2.3). This part of the dose can be modified dramatically owing to soil factors that may modify the bioavailability, and hence toxicity, of the added TE (i.e., the "active" fraction). Because the use of SQSs in screening for potential risk is intended to be conservative (and therefore will screen out locations where risk is definitely negligible), this implies that the SQS should be set for a soil composition corresponding to maximal bioavailability of the TE (perhaps based on a chosen percentile distribution of soil properties). If this is done, soils where bioavailability is not maximal may be incorrectly identified as being at potential risk. Thus, if feasible, the correction of SQSs for the soil-specific influence of bioavailability is a desirable refinement of current approaches.

In human risk assessment, the term "bioavailability" is used to describe the processes controlling the partitioning of a TE into different forms within the human body, with the term "bioaccessibility" used for the processes controlling the rate and extent of uptake from the gut into the body. Alternatively, it is usual in ecological risk assessment to use the term bioavailability to describe the chemical processes controlling the rate and extent of TE uptake into the tissues of the organism. While there is potential for confusion between the current terminology for the human (Chapter 3) and ecological spheres, a key distinction is that in the ecological sphere, environmental conditions play a greater role in modifying TE bioavailability. A distinction needs to be made between bioavailable and toxic, as these are not necessarily the same (i.e., TEs that are bioavailable need not be toxic, but the opposite is true).

A given level of toxic effect (e.g., 50% inhibition of reproduction of the earthworm *Eisenia fetida* after 28 days) of a TE on a soil organism can, in principle, be related to a given concentration of that TE at one or more sites of toxic action within the tissues of the organism. This concentration at the site of action should be independent of soil properties, assuming that the soil chemistry is within the range required for normal physiological function of the organism. Therefore significant variation in a toxic endpoint (as total or added TE) across soils implies a relationship between the amount of total or added TE and its concentration at the site(s) of toxic action that is mediated by the variability in relevant soil chemical properties. Understanding how TE toxicity appears to vary across soils requires an understanding of how the soil chemistry affects 1) the chemistry (speciation) of the TE and 2) the relationship between the speciation of the TE and the concentration at the site of toxic action within the tissues of the organism.

In setting out the state of the art in considering soil modification of the effective dose we will first consider the current state of knowledge regarding the chemical speciation and bioavailability of TEs in soils and then consider how such knowledge has been applied in seeking models and/or measurements for quantifying the effective dose. We will assess the effectiveness of current approaches to quantifying the effective dose and their application to the derivation of SQSs and related risk assessment. We will then highlight gaps in current knowledge and data and make recommendations for further research to improve the ways in which the factors that modify the effective dose of TEs can be better incorporated into methods for SQS derivation.

2.3 CONCEPTUAL MODEL OF THE SOIL-ORGANISM SYSTEM

Figure 2.6 illustrates the main forms in which a TE would be expected to be encountered in a soil-organism system and the main pathways of chemical transformation and transfer into organisms. Figure 2.6 is intended to show pools and transformation pathways of TEs in a broad sense; thus it is not necessarily applicable to all systems (e.g., the route of TE uptake via a digestive system is clearly only available to higher animals such as invertebrates) and there are important differences in 1) the significance of different chemical forms for different TEs and 2) the relative importance of different uptake routes and TE pathways for different soil organisms. Thus a key aspect of the concept of bioavailability is that it is highly context specific, particularly with respect to the specific organism(s) under consideration.

The distribution of a TE among its possible chemical forms is a complex function of the intrinsic chemical behavior of that element and its interactions with other soil constituents:

- the soil pH and concentrations of other major and TEs (e.g., Na, Mg, Al, K, Ca, Fe, SO_4, PO_4);
- the amounts of soil solid phases able to bind the element (e.g., organic matter, Fe and Mn oxides, clays);

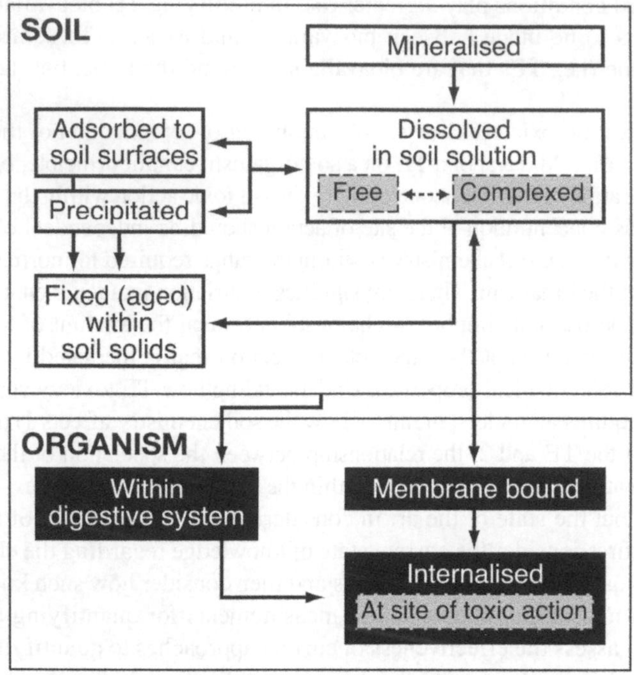

FIGURE 2.6 Simplified scheme for the main forms of a TE within the soil-organism system. White boxes represent TE pools in soil, black boxes represent locations within the organism. Arrows represent putative pathways of transformation among forms, assuming uptake of TE both directly from the soil solution and by uptake from ingested soil or food material.

- concentrations of ligands in the soil solution (e.g., dissolved organic matter); and
- for the removal of TE from the bioavailable pool, the time since addition.

The relative importance of these interactions will depend on the TE. For example, variability in C_a concentration across soils can influence the speciation of cationic TEs, but this is likely to be more pronounced for elements that bind relatively weakly, e.g., Cd, than those that bind strongly, e.g., Cu. The degree to which an element may form a precipitate will depend strongly on soil composition; for example, the formation of phosphate precipitates of Pb such as chloropyromorphite will depend strongly on soil phosphate levels.

Models are available that can describe the distribution of TE among different soil pools under the assumption of chemical equilibrium; such models are typically used to simulate TE speciation in the soil solution (e.g., Weng et al. 2002) or the distribution of TEs between the soil solids and the soil solution (e.g., Dijkstra et al. 2004; Gustafsson, 2006). Kinetic modeling of these systems has been much less researched, but some research into the kinetics of desorption from soil solids has been done (e.g., Barrow et al. 1989; Amacher et al. 1998; Shi et al. 2008). The slower processes of

TE weathering and fixation have been studied for some time (Reddy and Perkins 1974; Bruemmer et al. 1988), but models to predict bioavailability at toxicity-based concentrations have only recently been developed (e.g., Ma et al. 2006; Wendling et al. 2009; Donner et al. 2010), following the realization of its importance as a means of reducing TEs toxicity over short timescales (e.g., 30 days).

Models for organism uptake of TEs can be considered in 2 main groups: models for the direct uptake of TE from the soil solution and models for the uptake of TE via ingestion. Here we shall focus on models for direct uptake from the soil solution because the majority of research has been focused on such models and because direct uptake is considered the key pathway of uptake for most soil-dwelling organisms, even soil-ingesting invertebrates (Scott-Fordsmand et al. 2004).

2.4 IMPLICATIONS FOR SETTING SOIL QUALITY STANDARDS

Incorporating the influence of soil modifying factors into the setting of SQSs must be done in the context of the type of SQS that is to be set. As we have already discussed, the influence of modifying factors will vary according to an organism's main pathway(s) of exposure to TE. The focus of SQS setting is usually on the protection of the ecological structure or function of soil-dwelling organisms. We shall therefore focus on efforts to account for the effects of modifying factors on toxicity of TEs to soil-dwelling organisms, although some consideration will be given to secondary poisoning of higher-order vertebrates.

The ways in which soil modifying factors could be taken into account in setting an SQS are contingent on how the standard is to be expressed, and, more importantly, on how soils are intended to be monitored for compliance with the standard. This latter point is central, because taking soil modifying factors into account could imply a move away from the expression of standards as a total soil concentration of the TE to an alternative system, e.g., the expression as a concentration of TE in the soil solution or a specific chemical extraction. The scientific and financial implications of such a move would need to be considered fully. Another possible consideration would be the desirability of using existing toxicity data in deriving the standard; such data are measured almost exclusively as total or added TE concentrations, and this would also need to be considered.

For an SQS that is to be expressed as a total concentration of the TE in the soil, an account needs to be taken of how the toxic endpoints used in the calculation of the SQS will vary with soil chemistry. Different toxicity tests are conducted in different soils. Thus the toxic effect concentration has 2 components: 1) the inherent sensitivity of the species being tested, and 2) the effect (if any) that soil properties have on toxicity. If there is a bias in the type of soil types used for the toxicity testing, then using such data in SSDs to calculate an SQS will also be biased. It is therefore preferable to remove the effect of soil properties or to at least minimize the effect as much as possible. This may, in principle, be done by 2 approaches (McLaughlin et al. 2000): use either empirical or mechanistic models to estimate the toxicity in one standard soil (a process called normalization) or to express the effective concentration in terms of a measured soil pool of the TE that correlates with the observed toxic effect.

2.5 MODELS OF TE UPTAKE AND TOXICITY TO SOIL ORGANISMS

Models for the bioavailability and effects of TEs on soil organisms can be considered under the broad headings of mechanistic and empirical. Mechanistic models seek to relate toxicity to the amount of TE present in the soil and/or soil solution system by explicitly considering both the chemical and biological mechanisms that control uptake and consequent toxicity of the TE. Current efforts to mechanistically model bioavailability to terrestrial organisms are focused on the Free Ion Activity Model (FIAM) and the terrestrial Biotic Ligand Model (tBLM) and related models for ionic uptake at the root surfaces of plants. Uptake is modeled as occurring at distinct uptake sites on the outer surface of the organism (e.g., the plasma membrane of a root or the dermis of an earthworm) directly from the aqueous phase. The chemical speciation of the TE in the aqueous phase is a key aspect of these models, as are interactions with other aqueous ions (e.g., H^+, Ca^{2+}) at the organism surface and the receptor sites.

In contrast to mechanistic models, empirical models describe relationships between soil composition and uptake and/or toxic effect using expressions that are as simple as possible. Although, in principle, empirical models could be derived by a data mining approach using as many soil compositional parameters as are available, in practice, the choice of parameters that are measured tends to be informed at least in part by a deeper chemical and/or biological understanding of the processes controlling bioavailability (cf. Figure 2.7) and issues associated with practicality of use. Thus the distinction between mechanistic and empirical modeling is not clear cut. As we shall see, the manner in which current mechanistic models have been applied in soils has sometimes had a notable empirical element.

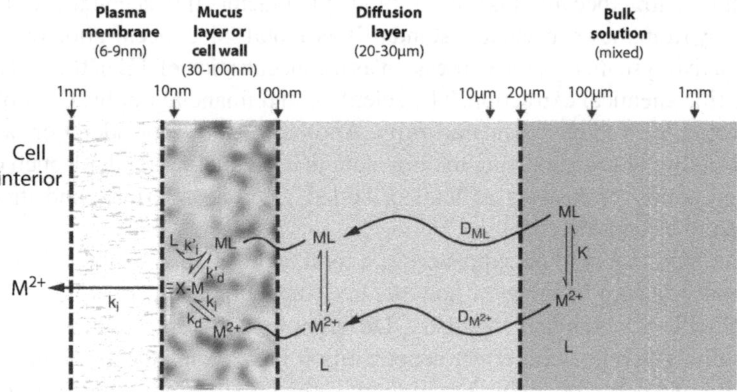

FIGURE 2.7 Conceptual model of metal-organism interactions (FIAM). Cationic free metal ions (M^{z+}) interact with ligands (L) in solution and diffuse at varying rates (D_{ML} or D_M) to the organism surface where they interact with surface ligands (EX). k_f, k_f' are rate constants for formation of the surface complex, k_d, k_d´ are rate constants for dissociation of the surface complex and k_{int} is the rate constant for transmembrane transport of the metal (reproduced from Campbell PGC Interactions between trace metals and aquatic organisms: a critique of the free ion activity model. In Metal Speciation and Bioavailability in Aquatic Systems. (Eds A Tessier and DR Turner), Copyright (1995), with permission from John Wiley and Sons).

2.6 MECHANISTIC MODELS

2.6.1 The Free Ion Activity Model and Biotic Ligand Model

The FIAM predates the tBLM. In fact, the tBLM is an enhanced version of the FIAM. Both postulate that direct uptake of a TE from the soil solution surrounding an organism can be directly related to the concentration or activity of its free ionic form in solution (Campbell 1995). It postulates a relationship between free and organism-bound TE that is directly analogous to a chemical equilibrium between free and complexed forms of a TE:

$$\text{TE} + \text{Org} \leftrightarrow \text{TE-Org};\ K = [\text{TE-Org}]/[\text{TE}][\text{Org}]$$

where Org is the receptor site, TE is the free ionic form of the TE in the soil solution, [TE-Org] is the concentration bound at a membrane receptor site Org, and K is a constant describing the relationship between receptor-bound and free TE [TE]. From the "equilibrium" expression we can write

$$[\text{TE-Org}] = K[\text{TE}][\text{Org}]$$

and if we assume that the concentration of receptor-bound element is much lower than the total concentration of receptor ([TE-Org] $\lll$ [Org] + [TE-Org]), then we can say that [TE-Org] is proportional to [TE]. We then further assume that the rate of element transfer to the site of toxic action, following binding to the receptor, is slow in comparison with the initial binding of the element from solution, such that the toxic effect can be related only to [TE-Org] and thus to [TE]. There is a critical, although sometimes overlooked, proviso to the theory that the element in question is not necessarily the only element that may bind to the receptor. Thus the theory that toxicity is proportional to free ion activity only holds where concentrations of other elements that bind to the receptor are constant. Where the concentrations of other binding elements change, it is necessary to consider their effects via competitive binding equations. Such competitive binding is the basis of the Biotic Ligand Model (BLM). The BLM postulates that the biological receptor is not only able to bind the TE of interest but also other elements in a competitive manner analogous to the competitive binding of ions to a chemical ligand. For currently developed BLMs for aquatic and soil organisms that relate to cationic TEs such as Cu(II) and Zn(II), competing ions invoked have included H^+, Na^+, Mg^{2+}, K^+, and Ca^{2+} (e.g., de Schamphelaere and Janssen 2002; Borgmann et al. 2005; Steenbergen et al. 2005). In the BLM the key parameter is the concentration of receptor (biotic ligand)-bound element [TE-Org], which for a given level of toxic effect on a given organism is assumed constant and independent of the chemistry of the ambient media. The BLM was initially developed to account for the acute effects of cationic TEs on the functioning of fish gills but has since been developed for acute and chronic effects on aquatic invertebrates also (e.g., de Schamphelaere and Janssen 2002; Borgmann et al. 2005). A key aspect of the FIAM and BLM is that they assume chemical equilibrium in the external medium (i.e., the soil solution). A chemical speciation model is thus an essential submodel within the FIAM/BLM, although, in principle, if free ion activities of all the relevant competing cations in the soil solution could be measured, it would be possible to apply the biotic ligand binding model without a chemical speciation model.

There is a reasonable body of research in the literature on the mechanisms of the uptake and toxicity of TEs by plants (McLaughlin 2001b) because there is an ongoing interest into the mechanisms of the plant uptake of TEs by crops. This body of research is largely driven by the need to understand the mechanisms of plant nutrient uptake, although there have been studies on the toxicity of TEs also, e.g., on selenate toxicity (Kinraide 2003) and on Cu, Zn, and Al toxicity (Kinraide et al. 2004). There is a closely linked body of work on the mechanistic binding of TEs to plant root surfaces (e.g., Yermiyahu et al. 1997; Rufyikiri et al. 2003) without the explicit consideration of resulting toxic effects. In addition, recent interest in developing the BLM for soil organisms has stimulated new research. Because of the possibility of studying TE interactions with plant roots in readily controllable nutrient solutions, work has been able to focus on a number of issues, such as the influence of small organic acids on cation uptake by roots (e.g., Parker et al. 2001). Root uptake of TEs has been explicitly modeled by a competitive ligand-binding type model, analogous to the BLM (Weng et al. 2002). Some researchers have developed root binding models that incorporate the electrostatic potential at the plant root plasma membrane surface. The electrostatic potential is determined by the ionic composition of the ambient solution and itself influences the activities of ions adjacent to the root-solution interface, thus exerting important effects on TE interaction with the root surface. For example, Gimmler et al. (2001) studied the effect of changing pH on the uptake of ions by the filamentous fungus *Bispora* sp. Decreasing pH decreased the uptake of most cations but increased the uptake of anions. Increasing pH has the opposite effects. The authors interpreted the variations in ion uptake as being strongly dependent on the surface electrical charge of the fungal filaments. Kinraide (2003) studied the effects of selenate on wheat (*Triticum aestivum*) and found that increasing concentrations of H^+, Mg^{2+}, Ca^{2+}, and Sr^{2+} in the rooting medium all adversely affected the growth of roots and increased the concentration of Se within the root tissues. He interpreted these results in terms of the electrostatic potential of the plasma membrane at the root surface; increased cation concentration lowered the potential (i.e., decreased the negative charge at the root surface) and thus increased the activity of selenate at the membrane.

Another notable aspect of research into toxicity of TEs to plants has been the study of the effects of binding ligands (both organic and inorganic) on root uptake and toxicity. Parker et al. (2001) grew wheat (*T. aestivum*) in the presence of Cu or Zn and organic acids malonate, malate, and citrate. Malonate and malate ameliorated Cu toxicity but not to the extent predicted by the FIAM; in other words, the metal-organic complexes were in some manner contributing to the toxicity. This effect was not seen when plants were exposed to Cu in the presence of citrate nor was it observed when plants were exposed to Zn in the presence of any of the organic ligands. Smolders, McLaughlin, and coworkers (Smolders and McLaughlin 1996; Smolders et al. 1998) observed that Cd uptake by Swiss chard (*Beta vulgaris*) from nutrient solutions and from soil was enhanced by the presence of chloride. Antunes and Hale (2006) studied the influence of a strong chelating agent (NTA) on the apparent binding strength of Cu to roots of wheat (*T. aestivum*) in nutrient solutions. The apparent binding strength of Cu in the presence of NTA was over 3 orders of magnitude higher than in the absence of NTA. The authors considered that direct uptake of Cu-NTA complexes was unlikely and suggested that NTA enhanced Cu uptake by increasing the rate of metal supply from bulk solution to the root surface. McLaughlin et al. (1997) grew lettuce (*Lactuca sativa*) and found that the uptake of Cd and Zn was increased (at constant free metal activity) in the presence of organic ligands. Degryse et al. (2006a) also found that uptake of Cd by spinach (*Spinacia oleracea*) was increased (at constant free Cd^{2+} activity) in the presence of synthetic ligands capable of forming labile complexes with the metal. Degryse et al. (2006b) studied the uptake of Cd by spinach (*S. oleracea*) and wheat (*T. aestivum)* in the presence and absence of the complexing agents HEDTA, CDTA, and NTA and demonstrated a diffusion limitation of root TE uptake in the absence of complexing agents. Thus the uptake of TEs by plants appears in some circumstances to violate one of the key assumptions of the FIAM, namely, that uptake is not limited by diffusion from the bulk solution to the organism-solution interface. What is not yet clear, however, is whether this is the case in intact soils, where roots may exude ligands to perform the buffering function carried out by the complexing agents in these experiments.

Antunes et al. (2006) reviewed the prospects for the development of the BLM for application to plants. They considered that there were several key challenges involved in BLM development, particularly, the reliable measurement of free ion activities and ligand concentrations in the rhizosphere, the identification of the organism "ligands" associated with toxicity, and the possible need to incorporate kinetic dissolution of metal-ligand complexes as sources of free ion. Notwithstanding these research needs, research into BLMs for plants is ongoing. Thakali et al. (2006a, 2006b) applied a BLM to data sets on the chronic toxicity of Cu and Ni to barley (*Hordeum vulgare)* and tomato (*Lycopersicon esculentum*) in 11 (Cu) and 8 (Ni) noncalcareous soils from across Europe. The BLM consistently provided a better fit to the data than did using the added soil metal or the free metal ion as the measure of concentration. It is notable that this study fitted the BLM without explicit consideration of the nature of the biotic ligand and in the absence of any measurement of bioaccumulated metal that could be related to the amount of Cu bound to any such ligand. Antunes et al. (2007) exposed barley (*H. vulgare*) in nutrient solutions to Cu in the presence of NTA and EDTA as buffering agents and varying pH and concentrations of Na, K, and Ca. Low-affinity root binding ligands were identified as the important binders of Cu because high-affinity ligands were largely occupied (~99%) already at concentrations of Cu associated with toxic effects. The amounts of Cu bound to the low-affinity root ligands correlated excellently with root growth inhibition (Figure 2.8), and solution concentrations inhibiting root growth by 50% (IC50s) calculated from a BLM were statistically indistinguishable from measured values. Lock et al. (2007a) developed a BLM for the toxic effects of Co^{2+} on barley in nutrient solutions, finding effects of Mg^{2+} and K^{+}, but not H^{+}, Na^{+}, or Ca^{2+}. This model was able to predict the toxic concentrations of Co^{2+} in 2 soils within a factor of 4. This is not as good a prediction as might be expected on the basis of previous work on aquatic organisms; the authors speculated that this might be due to differences in the solution chemistry close to the root surface (i.e., in the rhizosphere) between the

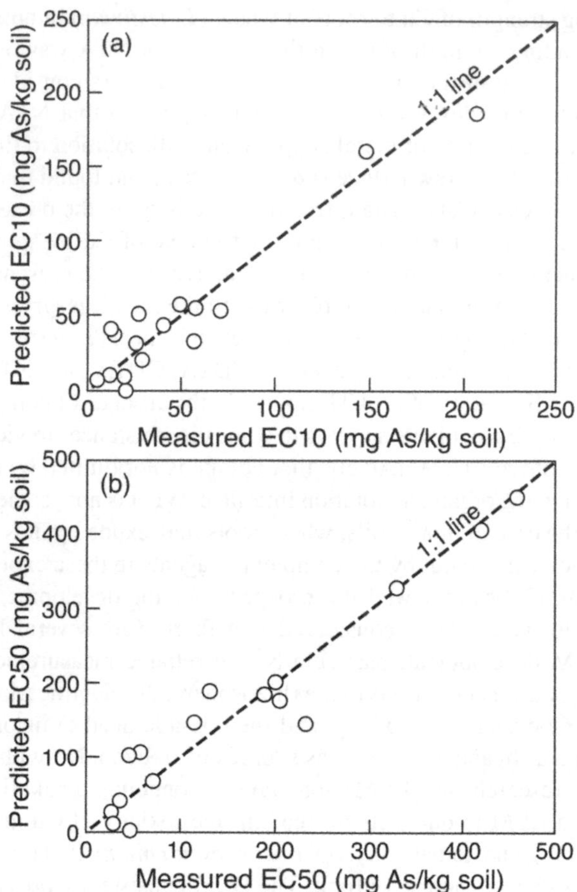

FIGURE 2.8 Relationship between measured and predicted (a) 10% effect concentration (EC10) and (b) 50% effect concentration (EC50) of added As that caused the specified percentage inhibition on root elongation of *Hordeum vulgare* (reproduced from Environmental Toxicology and Chemistry 25, Song et al., Influence of soil properties and aging on arsenic phytotoxicity, Copyright (2006), with permission from the Society of Environmental Toxicology and Chemistry).

solution only and soil experiments, as was suggested by Antunes et al. (2006). It is notable that the nutrient solutions used did not contain metal buffers, unlike those used by Antunes et al. (2007).

Some workers have also applied the tBLM concept to invertebrates and microbes. Steenbergen et al. (2005) developed a tBLM for acute toxicity of Cu to the earthworm species *Aporrectodea caliginosa*. The animals were exposed in a purpose-built system designed to limit exposure via diet. The free Cu^{2+} ion was measured using an ion selective electrode. Experiments showed that toxicity of Cu when expressed as the free Cu^{2+} was a function of the activities of the H^+ and Na^+ ions. However, Mg^{2+} and Ca^{2+} had inconsistent effects on Cu toxicity. The authors also identified a protective effect of DOC on toxicity, greater than could be explained by the complexation of Cu^{2+}. A BLM incorporating the protective effects of H^+ and Na^+ on Cu toxicity was developed and found to predict toxicity well, although a simple linear regression $p(LC50_{Cu2+}) = f(pH, p_{Na})$ was found to provide a superior prediction. As with the work of Thakali and coworkers, this study is notable in that the BLM was fitted directly to the effects data without consideration of TE concentration at the site of action. Koster et al. (2006) undertook a validation exercise for the Steenbergen et al. (2005) model by measuring the 28-day mortality of *A. caliginosa* in field-contaminated soils. The basic premise of the model was validated, although there was an underestimation of toxicity, likely due to the presence of other toxic TEs besides Cu.

Thakali et al. (2006a, 2006b) developed tBLMs for the chronic toxicity of Cu and Ni to the springtail *Folsomia candida* and the earthworm *Eisenia fetida*, in 11 (Cu) and 8 (Ni) noncalcareous soils from across Europe. The tBLM consistently provided a better fit to the data than did using the total soil metal or the free metal ion as the measure of concentration.

Lock et al. (2006) developed a tBLM for the acute toxicity (14-day mortality) of Co^{2+} to the potworm *Enchytraeus albidus* in sand infused with nutrient solutions of varying chemical composition. Effects of H^+, Mg^{2+}, and Ca^{2+} (but not Na^+) on Co^{2+} toxicity (expressed as the Co^{2+} activity) were found and a BLM developed on this basis. This model was tested by prediction of LC50s in a standard artificial soil and a field soil; in both cases the prediction was within a factor of 2 of the observed LC50.

Mertens et al. (2007a) compared the effect of Zn on nitrification rates in soil and soilless culture and found protective effects of H^+, Ca^{2+}, and Mg^{2+} against Zn^{2+} toxicity. They developed a tBLM and a Freundlich isotherm-type binding model to describe their results. Both models performed well within the range of calibration, and both could describe the effects of Zn^{2+} in soil and soilless culture well with the same set of parameters. This provided a strong indication that the microorganisms responsible for the nitrification process were exposed to Zn via competitive uptake with other solution cations at their cell surfaces.

Van Gestel and Koolhaas (2004) measured the toxicity of Cd to the collembolan *F. candida* in 7 soils (including soils of otherwise identical composition whose initial pH was amended). Total and free Cd ions were measured in water-extractable and porewater fractions. Toxic endpoints (EC50s) for reproduction and growth varied by a factor of 20 when expressed as total soil Cd, by 4.5 when expressed as water-extractable Cd, and by 8 when expressed as water-extractable free ion. When expressed as tissue Cd, however, the EC50s varied by a factor of only 1.7. The concentration of tissue Cd could be predicted very well ($R^2 = 0.74$) from either porewater or water-extractable Cd^{2+} and pH using an expression described by the authors as an extended Langmuir isotherm but which is conceptually identical to the BLM. The predicted tissue concentrations could be used to make an excellent prediction of the effect of Cd on animal growth ($R^2 = 0.89$). Modeled tissue Cd was a superior concentration measure for growth effects compared with any other measure of concentration and could be related back to a function of free ionic Cd and pH in soil solution. This is a key study among existing tBLM-type work for soil organisms, as it demonstrated that the link between soil solution chemistry (pH and free Cd^{2+}) and effect could be explained by accumulation of Cd in the particular organism used.

It is evident from the above studies that much progress has been made in describing both solution speciation of TEs in growth media and also in quantifying the competitive ion effects that underpin the concept of the tBLM. However, inconsistencies in competitive ion effects on TE toxicity indicate that more work is required in this area to understand the mechanisms responsible, paying special attention to measurements and confirmation of TE solution speciation.

2.6.2 Models Using Adsorption Isotherms

Some workers have developed models that are conceptually similar to the BLM in that chemical speciation and biological uptake are considered as separate processes, while using a different model for the biouptake component. Plette et al. (1999) modeled the binding of Cd and Cu both to soils and to biota (fungal mycelia, yeast cells, maize root cell walls) and demonstrated it could be described by a single type of model, the extended Freundlich isotherm:

$$Q = K_s \cdot a_M^{x} \cdot a_H^{y}$$

where Q is the amount of TE bound; a_M and a_H are the solution activities of the free ionic TE and the proton, respectively; and K_s, x, and y are empirical constants. This example considers only proton competition; but competition from other ions can also be modeled. Plette et al. (1999) used their models to explain how trends in bioavailability of Cu and Cd were ambient medium (soil/water) dependent.

2.6.3 The Free Ion Approach

The free ion approach separates chemical and biouptake processes and describes each in an empirical manner, which allows for the derivation of a single unifying expression. In the study of Lofts et al. (2004), this allowed the expression of the critical labile soil metal as a function of pH, soil organic matter content, and labile soil metal (the pool of metal including that in solution and adsorbed to soil surfaces). The method takes a mechanistic view of chemical speciation in the soil solution and couples this with a simple empirical expression by which toxicity is assumed to be a function of the free ionic form of the TE and the soil solution pH:

$$\log [TE]_{free,\,tox} = \alpha \times pH_{ss} + \gamma$$

where $[TE]_{free,\,tox}$ is the concentration of the free TE exerting a given level of toxic effect, pH_{ss} is the soil solution pH, and α and γ are fitting parameters. The effect of pH on toxicity is interpreted as being "protection" against the toxicity of the free ion. The approach is related to the use of the extended Freundlich isotherm (Plette et al. 1999; Mertens et al. 2007b) because the isotherm equation can be rearranged to an expression equivalent to that above, but the free ion approach explicitly does not include organism-bound TE. Concentrations of other soil solution cations could be incorporated into the expression, although this was not done by Lofts et al. (2004);

instead, an assumption was made that concentrations of other cations would covary with pH, thus removing the need to consider them explicitly. The expression was applied to chronic toxicity data from the literature for plants, invertebrates, and microbial processes. By regressing the entire data set for each metal against pH_{ss} and assuming the regression residuals to represent the distribution of organism sensitivities, expressions for the concentration of free ion protecting 95% of all species or processes were derived. The toxicity of the free metal ion was shown to depend strongly on pH; that has been confirmed more recently by Oorts et al. (2006a) for Cu effects on microbial processes, by Broos et al. (2007a) for Zn effects on microbial processes, and by Rooney et al. (2007) for Ni effects on barley (but not tomato).

2.7 EMPIRICAL TOXICITY MODELS

Empirical toxicity models seek to quantify the variability of TE effects on an organism across soils of different compositions without any reference to the underlying mechanisms controlling such variability. In the absence of the need to be mechanistic, empirical models can be based purely on experimental observations, although, in practice, the choice of soil parameters investigated in deriving the model is likely to be informed by mechanistically based understanding of processes controlling the soil chemistry of the TE, e.g., the concentrations of key complexing materials within the soil. A notable difference between these empirical expressions and the tBLM is that the latter model differentiates between the chemical speciation of the TE in the soil and/or soil solution, and the process(es) by which the TE affects the organism.

In early work, Van Straalen and Denneman (1989) normalized toxicity data to a standard soil composition using equations intended to predict background TE concentrations from soil properties. Since then, and particularly since about 2000, there have been a number of studies that have used toxicity data from different soils to derive species-specific normalization relationships for selected plant and invertebrate species and for some microbial processes (Table 2.2). The relationships largely refer to cationic TEs, although a recent study on As has also been done (Song et al. 2006).

As can be seen from Table 2.2, the soil chemistry variables predicting toxicity are almost always the pH, cation exchange capacity (CEC), soil organic carbon (OC), or clay content. This can be rationalized in that CEC, OC, and clay content are all measures of the TE binding capacity of the soil, while the soil pH both exerts a strong control on TE speciation and can act as a competing ion for binding to biotic ligands. Because the CEC is a soil property that depends on the amounts and nature of cation binding sites on the soil solids (including their pH-dependent binding behavior), it is notable that CEC is the best predictor of the variability in toxicity in a number of the studies. It is also notable that in the study of Song et al. (2006) on As(V), the only study available on the toxicity of an anionic TE, Mn and Fe oxides were found to be important predictors of the variability of As(V) toxicity. This is broadly consistent with the fact that mineral oxide surfaces are known to be potentially important complexants of As(V) (e.g., Dzombak and Morel 1990).

TABLE 2.2
Summary of published studies relating toxicity expressed as total soil metal to soil chemistry parameters

TE	Organism/process	No. soils	Soil composition			Significant parameter(s) in best linear regression	Independent variable (toxic endpoint)	R^2	Reference
			pH	C_{org} (%)	Clay (%)				
Ni	Potential nitrification rate	16	3.6–7.7	0.31–33.1	1–55	log CEC[1]	log EC50	0.70	Euras 2006
Ni	Glucose-induced respiration	16	3.6–7.7	0.31–33.1	1–55	log CEC[1]	log EC50	0.82	Euras 2006
Ni	Maize residue mineralization	16	3.6–7.7	0.31–33.1	1–55	log CEC[1]	log EC50	0.55	Euras 2006
Ni	*Hordeum vulgare*	16	3.6–7.7	0.25–33.1	0.4–55	log CEC	log EC50	0.90	Rooney et al. 2007
Ni	*Lycopersicon esculentum*	16	3.6–7.7	0.25–33.1	0.4–55	log CEC or log Ca_{exch}^2	log EC50	0.70	Rooney et al. 2007
Ni	*Folsomia candida*	16	3.6–7.7	0.25–33.1	0.4–55	log CEC	log EC50	0.84	Euras 2006
Ni	*Eisenia fetida*	16	3.6–7.7	0.25–33.1	0.4–55	log CEC	log EC50	0.75	Euras 2006
Co	*H. vulgare*	10	4.3–7.5	0.8–5.3	1–48	log Ca_{exch}^2, log Mg_{exch}^2	log EC50	0.92	Li et al. 2009
Co	*L. esculentum*	10	4.3–7.5	0.8–5.3	1–48	pH, log OC	log EC50	0.87	Li et al. 2009
Co	*Brassica napus*	10	4.3–7.5	0.8–5.3	1–48	log CEC, log oxide Mn[3], log C/N[4]	log EC50	0.86	Li et al. 2009
Co	*H. vulgare*	10	4.3–7.5	0.8–5.3	1–48	log Ca_{exch}^2	log EC50	0.86	Mico et al. 2008
Cu	*H. vulgare*	18	4.3–7.5	0.38–23.3	5–51	log Ca_{exch}, log oxide Fe[3]	log EC50	0.76	Rooney et al 2006
Cu	*L. esculentum*	18	4.3–7.5	0.38–23.3	5–51	log Ca_{exch}, log clay	log EC50	0.85	Rooney et al. 2006
Cu	*Triticum aestivum*	14	4.3–7.5	0.9–5.7	4–69	pH, log CEC	log EC50	0.91	Warne et al.. 2008a
Cu	*T. aestivum*	11	4.3–7.5	0.9–5.6	4–66	pH, log OC	log EC10	0.80	Warne et al. 2008b
Cu	*F. candida*	19	4.3–7.5	0.4–23.3	5–51	pH, log OC	log EC50	0.75	Criel et al. 2008
Cu	*E. fetida*	19	4.3–7.5	0.4–23.3	5–51	log OC, log clay	log EC50	0.81	Criel et al. 2008

Cu	Potential nitrification rate	19	4.3–7.5	0.4–23.3	5–51	log CEC	log EC50	0.66	Oorts et al., 2006
Cu	Glucose-induced respiration	19	3.0–7.5	0.4–23.3	5–51	log Corg	log EC50	0.57	Oorts et al. 2006a
Cu	Maize residue mineralisation	19	3.0–7.5	0.4–23.3	5–51	pH	log EC20	0.52	Oorts et al. 2006a
Cu	Substrate-induced nitrification	12	4.0–7.6	0.9–5.7	4–66	pH	EC50	0.83	Broos et al. 2007a
Cu	Substrate-induced respiration	12	4.0–7.6	0.9–5.7	4–66	clay	EC50	0.38	Broos et al. 2007a
Zn	*Enchytraeus albidus*	12	4–7	0.73–54	’–	pH, OM5, pH*OM	LC50	–6	Lock et al. 2000
Zn	*E. fetida*	21	4–6.3	0.75–7.54	9.7–20	pH, log CEC	log LC50	0.70	Lock et al. 2001
Zn	*E. fetida and E. andrei*	20	4–6.3	0.75–7.54	9.7–20	pH, log CEC	log EC50	0.69	Lock et al. 2001
Zn	*F. candida*	16	4.7–6.3	0.75–54	1.9–20	pH, log CEC	log LC50	0.87	Lock et al. 2001
Zn	*F. candida*	26	4.7–6.3	0.75–54	1.9–20	pH, log CEC	log EC50	0.38	Lock et al. 2001
Zn	*T. aestivum*	14	4.0–7.6	0.9–5.7	4–69	log CEC	log EC50	0.75	Warne et al. 2008a
Zn	*T. aestivum*	11	4.0–7.6	0.9–5.6	4–66	log CEC, background Zn	log EC50	0.89	Warne et al. 2008b
Zn	Substrate-induced nitrification	12	4.0–7.6	0.9–5.7	4–66	pH	log EC50	0.55	Broos et al. 2007a
Zn	Substrate-induced respiration	12	4.0–7.6	0.9–5.7	4–66	CEC, background Zn	EC50	0.46	Broos et al. 2007a
Cd	*E. albidus*	12	4–7	0.73–54	’–	pH, OM5, pH*OM	LC50	–6	Lock et al. 2000
As(V)	*H. vulgare*	16	3.4–7.6	0.4–23.3	7–51	clay, oxide Mn[3], oxide Fe[3]	EC50	0.91	Song et al. 2006

1 Only the results of simple linear regressions were given.
2 Exchangeable soil calcium (Ca) or magnesium (Mg).
3 Measured by extraction with oxalate.
4 Assuming organic matter contains 50% C.
5 Organic matter. Expressions were also derived using CEC instead of OM with very similar results.
6 Not quoted.

Figure 2.8 illustrates the relationship between measured As(V) toxicity in a range of European soils (Song et al. 2006) and the toxicity predicted by single and multiple linear regression equations against soil properties. The multiple regression approach provides reasonably good predictions of the As(V) toxicity across a 17-fold variability in the measured EC50 expressed as added As(V).

$$EC10 = 0.11(\text{oxalate-extractable Mn}) + 1.03(\%\text{clay}) - 9.25$$

$$EC50 = 0.21(\text{oxalate-extractable Mn}) + 0.016(\text{oxalate-extractable Fe}) + 4.29(\%\text{clay}) - 48.2$$

The possibility of using empirical relationships to adjust toxic endpoints for different soil compositions raises some practical issues. An empirical relationship can only be used on a particular endpoint if the soil property or properties that are used in the relationship have been measured on the soil used for toxicity testing. This means that the best practical empirical relationship is not necessarily that which gives the best fit to the calibration data but is rather the relationship that incorporates soil composition parameters that can routinely and easily be measured. Additionally, the best practical relationship is likely to be based on a limited number of soil parameters. The choice of relationship is therefore a balance between scientific justification and practicality. Also, owing to the time and resources necessary to develop robust empirical relationships for a given soil species, it may be necessary to use empirical relationships derived for one species or process to normalize toxic endpoints for other species and processes. For example, models developed for a single plant species (e.g., barley) may need to be extrapolated to ecotoxicity data for other plant species in the database for a given TE. In this situation, the choice of relationship to be used for a certain species or process needs to be carefully evaluated and justified as far as possible. Some guidance on extrapolating such empirical normalization relationships across species has been developed (Denmark 2008; ECHA, 2008; Warne et al. 2009).

2.8 DIRECT MEASUREMENT OF TE POOLS

Direct measurement of TE pools to take account of modifying factors is an attractive alternative to modeling because it may be a simpler and more cost-effective way of accounting for bioavailability. However, this method precludes prediction of toxicity thresholds from easily measured soil properties often held in existing spatial soil data sets (e.g., total metal concentration, clay content, pH, etc.). Direct measurement of an available pool of TE in soil is more often used as a tool in a second-tier risk assessment, where additional site-specific information is required.

Several reviews have been published of techniques to measure availability of TEs in soils (Lebourg et al. 1996; McLaughlin et al. 2000; Menzies et al. 2007; Peijnenburg et al. 2007). There are 10 classes of extraction for the quantification of TE pools in soils:

1) Weak extractants: water (this can include extraction of soil solution), water salt solutions, e.g., $CaCl_2$, $Ca(NO_3)_2$, ammonium acetate, Mg salts, $BaCl_2$;

2) Reductive extractants: sodium ascorbate, hydroxylamine-HCl, sodium dithionite (hydrosulphite) $Na_2S_2O_4$;
3) Competitive extractants relying on strong desorption: phosphate reagents for As, Mo, Se, etc. (e.g., Song et al. 2006; Sarkar et al. 2008);
4) Weak acids: diluted solutions of acids, especially acetic or citric acid;
5) Strong complexing agents (chelating agents), such as EDTA, DTPA (sometimes in combination with triethylamine and ascorbic acid), and NTA;
6) Combined salt-acid extractants: ammonium oxalate-oxalic acid, sodium acetate-acetic acid, HNO_3 + NH_4F + HAc + NH_4NO_3 + EDTA (Mehlich III), and many others;
7) Dilute strong acids: HNO_3, HCl, "double acid extraction" HCl+H_2SO_4, also named Mehlich I;
8) Concentrated strong acids: HNO_3, HCl, HNO_3 + HF, "aqua regia" (concentrated HNO_3 + HCl), Fleischmann acid.
9) Measures of free or labile ion: water or weak salt extraction of TE and analysis using ion selective electrodes (Cu), anodic stripping voltammetry, Donnan dialysis, or other methods;
10) Diffusion/speciation methods: selective extraction of free ion and labile ions using a an ion sink, e.g., diffusive gradients in thin films (DGT) (Zhang et al. 2001).

The methods listed above are largely applied with the aim of estimating the amount of TE in a specific "pool" within the soil, e.g., "total," "labile," "dissolved." Some are intended to alter the chemical composition of the soil by dissolving a defined solid phase or set of phases, with the intention of solubilizing associated TE; e.g., ammonium oxalate–oxalic acid extraction is commonly invoked as a means of dissolving iron oxide phases for analyzing their associated TEs. Broadly speaking, extractions with water or weak salt solutions are intended to isolate a "dissolved" fraction, strong complexing agents and strong acids are intended to extract "labile" TE (i.e., TE in solution and complexed to the surfaces of the soil solids, but not TE occluded into soil solids), while concentrated strong acids are intended to provide a measure of the "total" metal by causing at least the partial dissolution of the solid phase. In practice, only an extremely strong extractant, e.g., HNO_3 + HF, can cause total dissolution of the soil solids, thus other strong extractions are sometimes referred to as measuring a "pseudo-total" metal pool.

In the context of taking inherent soil factors into account in SQSs, the goal of research into extraction methods is to identify a method that extracts a pool of TE from different soils that correlates with toxic effect better than does the measurement of the total metal (McLaughlin et al. 2000). The focus of such research has thus been toward the weaker extractants, i.e., those that are believed to extract "dissolved" TE. This is based on the idea that the concentration of TE in the dissolved phase better correlates with toxicity than a stronger extraction that removes TE from the solid phase also.

There is an extensive body of research on the utility of different extraction methods for both the uptake and toxicity of TEs to plants (e.g., McLaughlin et al. 2000; Schultz et al. 2004; Menzies et al. 2007) and soil invertebrates (e.g., Posthuma and Notenboom 1996; Conder and Lanno 2000; Vijver et al. 2001). A number of studies claim to have found that toxicity of cationic TEs is better explained by weak extraction than by total metal. For example, Posthuma and Notenboom (1996) found that the variability in performance of the earthworm *E. andrei* across a transect of field soils contaminated with Cd, Cu, Pb, and Zn from a smelter was described better by 0.01 *M* $CaCl_2$-extractable TE than by total, assuming that Zn was the dominant cause of toxicity. Nolan et al. (2005) found that plant uptake of TEs was best correlated to a different measured pool depending on the TE: soil solution was best for Pb and also for Zn, $CaCl_2$ extraction worked best for Cd, while for Cu the total TE worked best. Alternatively, Vijver et al. (2001) found that porewater and 0.01 *M* $CaCl_2$-extracted concentrations of the nonessential TEs Cd and Pb were not, by themselves, generally good predictors of their accumulation by *F. candida*; total soil concentrations were generally better predictors. The $CaCl_2$-extractable Cd was a good predictor of accumulation in combination with the soil contents of inorganic matter and Al oxide ($R^2 = 0.56$), while the porewater Cd was a good predictor in combination with the total soil Ca concentration ($R^2 = 0.52$). The total Pb concentration was a good predictor of accumulation in conjunction with the soil CEC ($R^2 = 0.54$), and reevaluation of the data also showed that $CaCl_2$-extractable and porewater Pb were slightly superior predictors in combination with pH ($R^2 = 0.63$ and 0.60, respectively) (Vijver et al. 2001).

Menzies et al. (2007) reviewed the use of both weak and strong extractants for predicting phytoavailability of cationic TEs. They concluded that, generally, extraction with weak salt solutions gave results that tended to correlate best with plant TE concentrations. They tentatively suggested that 0.01 M $CaCl_2$ was likely to prove the best extractant for cationic TEs, although they noted that across different studies there was not necessarily an agreement that a particular extractant was optimal for a metal and that some research (e.g., Gupta and Aten 1993) found that alternative extractions were superior to 0.01 M $CaCl_2$.

Some recent studies have taken the useful step of comparing specific extractions with other modeling approaches. As already noted, Van Gestel and Koolhaas (2004) showed that toxicity of Cd to *F. candida* could effectively be described by a BLM better than any other expression of concentration, including water-extractable and porewater Cd, but the range of soils used in this study was very limited (2 natural soils and one artificial soil at different pH values). Spurgeon et al. (2006) measured the reproduction rate of the earthworm *Lumbricus rubellus* along a transect of smelter-contaminated soil. Taking Zn to be the main contaminant exerting toxic effects, reproduction rates were compared to total soil Zn, 0.01 M $CaCl_2$-extractable Zn, Zn^{2+} free ion activity, and a function of $Zn^{2+}$and pH based on the free ion approach. The pH-adjusted free ion approach function gave the best relationship with toxicity, followed by total and $CaCl_2$-extractable Zn, with the free ion activity model giving the poorest relationship. Thakali and coworkers (Thakali et al. 2006a, 2006b) found that at tBLM was consistently superior to soil solution concentration in describing Cu and Ni toxicity to plants, invertebrates, and microbial processes in a range of European soils.

Weng et al. (2004) studied the toxicity of Ni to oats (*Avena sativa*) in 4 different soils, each of which was further amended to give 1 of 3 or 4 pH values. The EC50 values (as total Ni) derived from each soil were shown to depend mainly on the soil pH, with some further variability due to differences on the contents of Ni-binding solid phases. The authors noted that EC50s expressed as 0.01 M $CaCl_2$-extractable

Ni varied less than did the total Ni EC50s and suggested that this extraction might be of use in measuring Ni for risk assessment. Semenzin et al. (2007) studied the possibility of using 0.01 M $CaCl_2$-extractable TE concentrations in an SSD approach (see below) to derive an SQS. The toxicity of Ni to plants was chosen as an example. The authors found that the variability in toxicity (EC10) of Ni to *A. sativa* using the data of Weng et al. (2004) was much lower than other measures when expressed as 0.01 M $CaCl_2$-extractable Ni. They suggested that using a bioavailability-based measure of metal toxicity such as 0.01 M $CaCl_2$ extraction to derive toxic endpoints would produce more realistic and robust SSDs and SQSs.

A small amount of research has been done on selective extractions for anionic TEs. Song et al. (2006) studied the toxicity of arsenate (As(V)) to barley (*H. vulgare*) in 16 soils and extracted As(V) using 0.05 M ammonium sulphate ($(NH_4)_2SO_4$) and 0.05 M ammonium dihydrogen phosphate ($NH_4H_2PO_4$). The variability in toxic endpoint when expressed using concentrations measured with either extractant was somewhat lower than the variability in total As(V) endpoints (maximum/minimum EC50 data were 17.2 for total added As(V), 10.0 for $(NH_4)_2SO_4$-extractable As(V), and 14.1 for $NH_4H_2PO_4$-extractable As(V). The $(NH_4)_2SO_4$-extractable EC50s showed a significant ($p \leq 0.01$) relationship with pH, while the $NH_4H_2PO_4$-extractable EC50s could be related to oxalate-extractable Fe, Mn, and clay.

In recent years, specialized measurement devices have emerged as a new means of measuring TE pools in soils. Probably the most studied of these devices is the diffuse gradients in thin film (DGT) collector (Zhang and Davison 1995). Briefly, a DGT device consists of a layer of gel with a filter on the outside (i.e., in contact with the environment) and a sorbing medium on the inside. TEs diffuse through the filter and gel and are sorbed, setting up a diffusion gradient through the gel, which promotes further uptake. For cationic TEs the sorbant is usually Chelex resin; Fe oxide has been used as a sorbent for anionic TEs (Zhang et al. 1998). The dynamic manner in which DGT samples the environment means that when applied to soils, it can be used to characterize the rate at which metal is resupplied from the solid to the solution phase in response to the movement of metal into the sampler (Degryse et al. 2009). Zhang et al. (2001) measured DGT-available Cu in a series of soils and calculated the "effective concentration" (C_{EFF}), a conceptual pool of dissolved TE that accounts for the resupply of metal to the soil solution from the solid phase as it is depleted by the DGT device. The uptake of Cu by pepperwort (*Lepidium heterophyllum*) from a range of Cu-contaminated soils was measured and compared to C_{EFF}, soil solution Cu, free Cu^{2+}, and EDTA-extractable Cu. The best correlation with plant concentrations was seen for C_{EFF}. After this initial study, DGT has been used to evaluate the uptake of Cu, Zn, Pb, and U by plants (e.g., Nowack et al. 2004; Song et al. 2004; Zhang et al. 2004; Cornu and Denaix 2006; Zhao et al. 2006; Vandenhove et al. 2007). These studies have indicated that, generally, C_{EFF} is at least as good as standard wet extraction methods for measuring a pool of TE that correlates with plant uptake. Some authors have reached more ambiguous conclusions regarding the usefulness of DGT. Nolan et al. (2005), although finding that C_{EFF} was a good predictor of Cd, Pb, and Zn uptake by wheat, found that for Cu the total soil metal was a superior predictor. Koster et al. (2005) used DGT to measure uptake of Zn by several plant species and an isopod; although C_{EFF} was about as good as $CaCl_2$ extraction at predicting Zn uptake by plants, the authors concluded that DGT did not have an advantage over the more conventional extraction because it was a more complex method. Isopod uptake of Zn was not well predicted by DGT, possibly because the animals did not necessarily take their Zn up from the soil solution. Almas et al. (2006) found that C_{EFF} was a good predictor of spinach and ryegrass uptake of Cd and Zn when the metals were below toxic levels; under conditions of plant toxicity, C_{EFF} did not predict uptake well.

It is difficult to conclude that one particular extractant or type of extractant is superior to others in predicting TE bioaccumulation and toxicity across a range of soils or organism endpoints. Extractants that only extract a small fraction of the TE from soil often pose problems in terms of chemical analysis (due to the low resultant solution concentrations), which leads to a lack of method robustness across laboratories. Furthermore, given that accumulation and effects of TEs are affected by other competing ions in the soil, it is perhaps too much to ask that a single extraction technique will perform well in all situations in predicting TE bioaccumulation and toxicity. Indeed, several authors have found that variability in toxicity endpoints based on partial extractants are more variable across soils than those based on total concentrations (Smolders et al. 2003, 2004; Oorts et al. 2006b; Zhao et al. 2006; Broos et al. 2007a; Warne et al. 2008b).

2.9 CONSIDERATION OF MODIFYING SOIL FACTORS IN SOIL QUALITY STANDARDS

In the previous sections we have focused on the chemical and biological interactions that control the direct uptake of TEs from soil solution and how models have been developed to describe these. This section will consider how modifying soil factors are currently implemented in SQSs and how we can extend and improve this implementation (this is discussed in more detail in Chapter 4). Some jurisdictions prescribe SQSs that either implicitly or explicitly incorporate the effects of soil modifying factors on TE toxicity, and these are summarized below. Unless otherwise referenced, this information is taken from the report of Carlon (2007).

Switzerland provides SQS values including pairs of values for Cd, Cu, F, Pb, and Zn expressed as pseudo-total (2 M HNO_3-extractable) and soluble (0.1 M $NaNO_3$-extractable) concentrations. A soil is considered to exceed the SQS if either or both of these concentrations are exceeded (Gupta et al. 1996). Austrian Standard ÖNOR M S 2088-2 provides generic SQS values for soil TEs related to risks of plant exposure; a simple scheme for soil-specific adjustment of these values is also provided. Additional screening values for NH_4NO_3-extractable TEs are also provided. Lithuania uses some SQS values expressed as mobile or bioavailable metal, originally derived by the former Soviet Union. Values for Co, Cu, Ni, and Zn are expressed as pH-buffered NH_4NO_3-extractable concentrations. Carlon (2007) states that these values have been modified according to national soil baseline values and taking into account European Union (EU) legislation and standards in neighboring countries but does not provide details. Carlon (2007) also states that values for water-soluble F and bioavailable Cr and F have been set but does not provide details. Germany has adopted

SQS values for As, Cu, Ni, and Zn that relate to growth impairment of cultivated plants, expressed as 1.0 M NH_4NO_3-extractable concentrations (BBodSchV 1999). Additionally, precautionary SQS values for the general protection of soil function for Cd, Cr, Cu, Hg, Ni, Pb, and Zn are differentiated according to soil texture. For Cd, Ni, Pb, and Zn, in acidic soils (pH < 5.0 for Pb, pH < 6.0 for the others) the SQS is lowered in some soils. Slovakia has adopted both these sets of values in its legislation. China has adopted SQS values for As, Cd, Cu, Hg, Ni, Pb, and Zn that are expressed as total metal and banded according to soil pH (separate values for pH < 6.5, pH 6.5 to 7.5, and pH > 7.5) (Ministry of Environmental Protection of the People's Republic of Chinas 1995).

There is clearly little commonality in current practice for the setting of SQSs that incorporate soil modifying factors. This is not surprising because SQSs can be developed for a specific purpose (e.g., precautionary values, intervention values) within the risk assessment process and have different ecological targets (e.g., generalized protection of soil functions, protection of crop plants). In part, also, the variety of approaches reflects the variety of scientific research that has taken place into soil modifying factors.

The consideration of modifying factors in the derivation of ecological SQSs for TEs has been increasingly recognized as a significant issue in recent years. A key focus of recent work has been the ongoing risk assessment process within the EU. Under this process, risk assessments for a number of compounds containing TEs of potential concern (e.g., Cd, Co, Cu, Ni, Pb, and Zn) are currently either completed or in progress. The recently (2008) completed risk assessment for Ni (Denmark 2008) included the consideration of modifying factors in deriving predicted no effect concentrations (PNECs) for the effects component of the assessment. Briefly, for a data-rich substance such as Ni, the PNEC is calculated as the 5th percentile (HC5) of the distribution of chronic endpoints (no observed effect concentrations [NOECs], which are considered to be equivalent to EC10 values) of terrestrial organisms (plants, invertebrates, microbes) to the substance. A further assessment factor may then be placed on the HC5 to give the final PNEC. As noted earlier, where endpoints for different organisms have been measured in soils of varying composition the optimal approach, if feasible, is to normalize each endpoint concentration to a "target" soil composition (e.g., a standard soil) prior to constructing the species sensitive distribution. In principle, this requires a normalization approach that has been validated for each organism or microbial process for which there is suitable endpoint data available. In practice, this is not currently feasible because of the large amount of experimental work that would be required to derive species- or process-specific normalization models; indeed, such work would likely render it unnecessary to use the existing EC10s or NOECs. Instead, for the Ni risk assessment, normalization models were generated for specific species and processes and tested for their ability to reduce the variability in EC10s or NOECs for similar species and processes. Chronic toxicity testing was done using a consistent set of 16 European soils covering a wide range of physicochemical properties. Tested organisms and microbial processes were barley and tomato (Rooney et al. 2007), *F. candida* and *E. fetida* (Criel et al. 2008), and potential nitrification rate (PNR), maize residue mineralization (MRM), and substrate-induced respiration (SIR) (Oorts et al. 2006a). Empirical normalization models were developed for each set of tests, based on CEC as an explanatory variable for the EC50, that is, EC50 = $a_{CEC} + b$.

To then apply these expressions to the effects data set, it was necessary to verify, that is, as far as possible that the variability in EC10s or NOECs would be reduced. It was necessary to relate specific normalization expressions to specific species and processes. This was done partly based on likely similarity of exposure and effects mechanisms and partly on considerations of generating a conservative guideline. Plant effects data, unless they related to tomato, were normalized using the expression for barley because this was considered to give slightly more conservative outcomes. Invertebrate effects data were normalized using the expression for *F. candida* for hard-bodied species and using the expression for *E. fetida* for soft-bodied species, based on expected differences in uptake mechanisms between these 2 groups. The PNR expression was used to normalize data relating to the soil nitrogen cycle, the MRM expression was used for data relating to respiration processes utilizing natural substrate, and the SIR expression was used for data relating to microbial biomass indicators. The ability of normalization to reduce variability in EC10s or NOECs for the same species or process in different soils was tested with all the available data; the results are shown in Figure 2.9 and indicate that either the normalization reduces the EC10s or NOEC variability or has little effect. The cases where the reduction in variability is small are those where the initial variability was also small. Thus the method appears to be a robust way of accounting for soil modifying factors in deriving SQSs, particularly, as it can reduce the bias in endpoints for species and processes other than those for which toxicity-soil composition relationships have been derived.

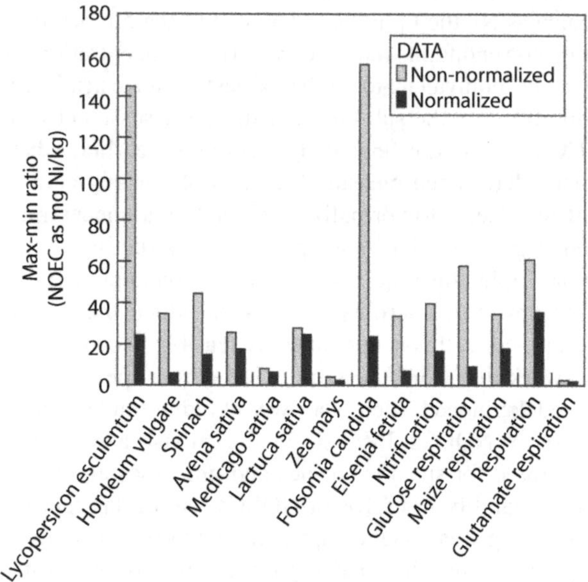

FIGURE 2.9 Intra-species variability (expressed as max/min ratios) of EC10/NOECs expressed as mg Ni kg^{-1} test medium and normalised, using chronic regression models.

As an example of the application of the method, a PNEC for Ni was calculated for 6 European soil scenarios. The PNEC varied between 8.6 and 161.4 mg kg^{-1} when taking the best SSD fit, demonstrating the importance of taking modifying factors into account. The methodology would be suitable, in principle, for calculating a generic SQS, e.g., for an "average" or "worst case" soil for broad screening, or alternatively could be applied on a purely location-by-location basis using local soil properties. A similar method has been used in the Cu voluntary risk assessment (European Copper Institute 2008). In the case of the Zn risk assessment (Netherlands 2004) the normalization of PNECs was rejected as a method; instead, a bioavailability-based correction was applied to the field exposure concentration (the predicted environmental concentration [PEC]). The correction factor used was the smallest of 3 possible correction factors, based on invertebrate, plant, and microbial toxicity test data. The correction factors for plants and invertebrates were based on log CEC, while the correction factor for microbial processes was based on the background Zn concentration. The correction of the PEC rather than the PNEC was justified on the basis that soil-type dependent correction is a laborious exercise because it requires recalculation of all NOECs in the database, but the correction could easily be applied to either contamineted of the risk characterization.

In parallel with the studies in Europe, a large set of experimental data for Cd, Cu, and Zn has been collected in Australian studies on microbial, plant, and food quality endpoints (McLaughlin et al. 2006; Broos et al. 2007a; Warne et al. 2008a, 2008b; Heemsbergen et al. 2009a, 2009b) collected on a wide range of soils spiked with metal salts. To date, these studies are unique in the range of soils used, the fact that soils were spiked and incubated under field conditions, and ecotoxicity endpoints determined under field conditions were compared with those performed in the laboratory. These studies have since been used to derive suggested SQSs for all 3 TEs in soils amended with biosolids (Heemsbergen et al. 2009a) and to derive proposed Australian SQSs for Cu and Zn in contaminated sites (Heemsbergen et al. 2009b; Warne et al. 2009).

2.9.1 Differences between Laboratory and Field Conditions in Ecotoxicity Studies

Soil toxicity tests form the basis for guidelines on the amount of a TE in soils that is considered to be safe for environmental health. Ideally, the toxicity of soils would be evaluated using field soils having a range of TE loadings sourced from historic contamination of a known loading—this is seldom the case. To overcome this problem, the ecotoxicity of TEs to soil organisms is assessed using soils that are artificially contaminated in the laboratory by adding TEs in the form of salts, oxides, or dissolved in a solution (e.g., Wilke et al. 2005; Oorts et al. 2006b). Recent studies have shown that soils that are spiked in this manner overestimate toxicity relative to soils with the same total concentration of TEs that have stabilized in the field (Smit and Van Gestel 1998; Lock and Janssen 2001; Fountain and Hopkin, 2004; Smolders et al. 2004). Conditions of artificially contaminated soils should be as close as possible to those of soils in the field to determine the toxic effects of TEs. In laboratory toxicity tests, however, spiking soil by adding TE salts could be problematic because the TE salts not only increase the total TE concentration in a soil but also affect other soil properties such as the ionic strength of the soil solution and pH (Speir et al. 1999; Stevens et al. 2003; Oorts et al. 2006b). The main problem in spiking soils with soluble TE salts is that the counter ion added, e.g., usually chloride (Cl), sulphate (SO_4), or nitrate (NO_3) for metallic cations such as Cd, Co, Cu, Ni, Pb, or Zn, may also have adverse effects on soil conditions or organism health. Chloride is widely known to be toxic to plants (Maas 1986), so that spiking soils with high rates of Cl salts may induce Cl toxicity as well as TE toxicity, or the effects may be combined. The high ionic strength of the soil solution in soils spiked with soluble TEs may also be deleterious to organism performance, akin to salt effects in irrigated soils (Richards 1954).

A further problem in comparing toxicity responses in field soils and laboratory-spiked soils is that of the source of contamination and TE aging. The source of the added TEs can have a large effect on the amount of dissolved or labile forms in the soil—highly insoluble additions will not increase solution TE concentrations or TE bioavailability markedly, a fact known for decades in fertilizer science (Mortvedt and Giordano 1969). Smolders (2000) reported that the effect of Ni oxide (which is insoluble) on nitrification rates was substantially lower than that of $NiSO_4$ and similar data have been reported for ZnO and tyre debris (which contains Zn) assessed against $ZnSO_4$ in soils (Smolders and Degryse 2002). These differences indicate a need to ensure that exposure of soil organisms to TEs is consistent among the toxicity data that are added to databases. This will assist in recognizing the linkages among the sources of added TEs, their distribution/bioavailability in soil, and their toxicity when comparing guidelines developed from laboratory toxicity tests (which are usually based on soluble TEs) to field-contaminated sites (which are typically characterized by insoluble forms of TEs). Hence toxicity may be markedly different in soils with the same total TE contamination due solely to source effects. Even where TEs are added as soluble salts, there are a number of slow continuing reactions in soil that reduce TE extractability, bioavailability, and toxicity (Crout et al. 2006; Ma et al. 2006; Oorts et al. 2007; Wendling et al. 2009; Donner et al. 2010). These issues are discussed in detail below.

2.10 EFFECTS OF SPIKING SOILS WITH SOLUBLE TE SALTS ON SOIL SOLUTION CHEMISTRY AND TOXICITY MEASUREMENTS

When solutions containing high levels of TEs are added to a soil, there are a number of rapid changes in soil properties. The added ions will displace adsorbed ions from the soil exchange sites as a new equilibrium between dissolved and adsorbed ions is being established. The net result is a sharp increase in ionic strength, a drop in pH and higher concentrations of dissolved cations (such as Ca, Mg, K, Cd, Cu, Ni, Pb, etc.) and anions (Cl, SO_4, NO_3, etc.). For cationic TEs, the lower pH (Speir et al. 1999) and higher ionic strength (Stevens et al. 2003) will cause a reduction in TE partitioning (K_d) for the same total metal concentration, hence increasing the soil solution TE concentration. This response can be clearly seen in the results of Schwertfeger and Hendershot (2008) (Figure 2.10). On days 1 to 5 the soil was leached with a

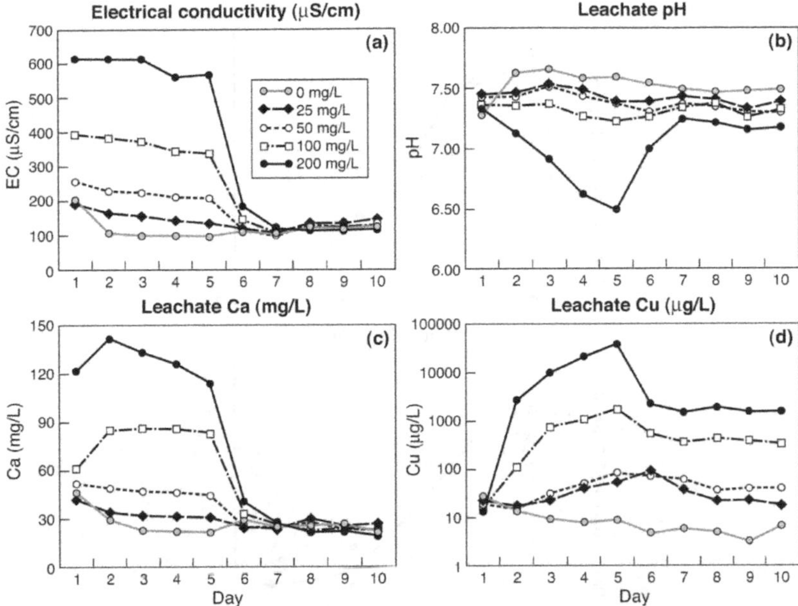

FIGURE 2.10 Changes after spiking soils with soluble Cu (at concentrations varying from 0 to 200 mg Cu/L) and subsequently leaching: a) electrical conductivity, b) pH, c) dissolved Ca and d) dissolved Cu.

Cu solution containing between 0 and 200 mg/L. During days 6 and 7 the soil was leached with water, whereas on days 8 to 10 a dilute Hoagland solution was applied to reestablish nutrient balance. The leaching with water and dilute Hoagland solution eliminated most of the artifacts caused by spiking the sample and provided a soil that had similar properties in terms of EC, pH, and dissolved Ca across the range of Cu values desired for ecotoxicity testing.

The adverse effects of counter ions will be greatest in soils with a high partition coefficient for the TE of interest, as high doses will be required to elicit a toxic response by the organism. Hence confounding effects of salt toxicity will be greatest in alkaline soils for metallic cations (Stevens et al. 2003).

Another artifact induced by spiking soils with soluble TE salts is the high concentrations of other cations in soil solution. Where toxicity of an ion at the organism membrane is affected by concentrations of other counter ions in solution, e.g., competitive effects of Ca, Mg, or other ions on TE toxicity, then performing experiments in soils that are spiked with soluble metal salts is clearly not representative of field soil conditions not only in terms of TE concentration but also concentrations of important cations and anions normally found in soil solution.

The net result of these differences in solution chemistry is that, for the same total TE concentration, toxicity is overexpressed in soils spiked with soluble TE salts. This has been demonstrated with both microbial and terrestrial plant ecotoxicity data. Smolders et al. (2004) examined soils in a transect away from galvanized electricity towers, with

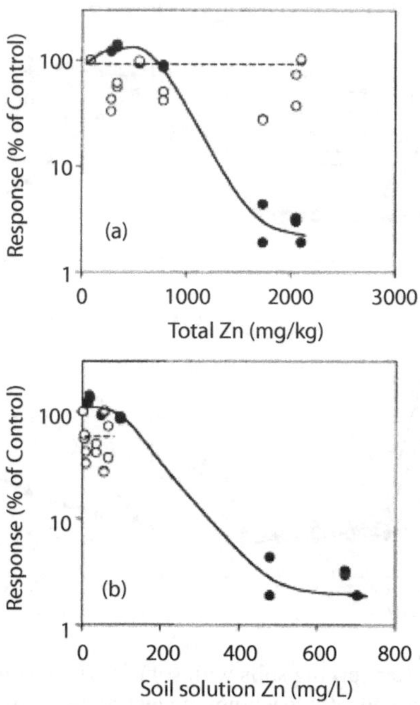

FIGURE 2.11 Potential nitrification rate in a field-contaminated soil and laboratory spiked soil at (a) equivalent total Zn concentrations and (b) soil solution Zn concentrations in the same soils (reproduced from Environmental Toxicology and Chemistry 23, Smolders et al., Soil properties affecting toxicity of zinc to soil microbial properties in laboratory-spiked and field-contaminated soils, Copyright (2004), with permission from the Society of Environmental Toxicology and Chemistry).

a gradient of Zn contamination clearly evident. Control soil (distant from the tower) was spiked in the laboratory to the same Zn concentration as the field gradient, and PNR and concentrations of Zn in soil solution measured. In the field contaminated soils, PNR was unaffected by Zn contamination, yet in the laboratory-spiked soils, PNR was reduced markedly by the same total Zn concentrations. The concentrations of Zn in soil solution were also markedly different (Figure 2.11).

2.11 MINIMIZING SPIKING-INDUCED ARTIFACTS IN THE LABORATORY

Immediately after spiking a soil with soluble TEs, there is a significant increase in ionic strength that can only be reduced by leaching. In the field, this leaching takes place naturally as long as precipitation exceeds evapotranspiration. In the laboratory, leaching can be used to reduce the high initial ionic strength of the soil solution to values similar to those prior to the addition of the TE spike. As yet, there is no standard method for treating soils to minimize the artifacts induced by spiking soils with highly soluble TE salts, although protocols for hazard assessment have been suggested (McLaughlin et al. 2002). Leaching and short-term aging are the 2 main treatments that have been assessed in laboratory-spiked soils to try to mimic TE bioavailability in field soils.

Several authors have leached soils with anywhere from 1 to 5 pore volumes of deionized water followed by time to equilibrate spiked TE with soil (Smit and Van Gestel 1998; Stevens et al. 2003; Bongers et al. 2004). Others have suggested submerging test soils in a synthetic soil solution (containing approximately 1.8 mM in total of base cations Ca, Mg, K, and Na) overnight followed by an addition of another pore volume of solution over saturated soils followed by gravimetric draining overnight (Oorts et al. 2006b). Only Stevens et al. (2003) mentions leaching soils until a particular ionic strength is achieved (i.e., electrical conductivity falls below a defined salinity threshold for the endpoint).

MacDonald et al. (2004a, 2004b) developed a procedure to remove the artifacts of soil sampling and drying so that soil solution chemistry would approach more closely that measured in the field. This approach has been modified to allow soils to be spiked and leached in the laboratory. Although the procedure tested to date is somewhat time consuming (10 days are required to spike and leach soil samples), the artifacts of spiking are largely removed and the soil nutrient balance is corrected by leaching with a nutrient solution to replace the nutrient ions lost during spiking (Schwertfeger and Hendershot 2008).

Oorts et al. (2006b) compared Cu toxicity thresholds for microbial processes (nitrification potential, glucose-induced respiration, and maize residue C mineralization) among laboratory-spiked soils that were 1) tested fresh, 2) leached by submersing spiked soil in a bucket of synthetic soil solution overnight followed by an addition of another pore volume of solution over saturated soils, or 3) aged for 6, 12, or 18 months outdoor under local weather conditions (780 mm annual precipitation) and allowed to drain freely. Leaching increased the partitioning coefficient of Cu in the soils during the toxicity test, resulting in higher toxicity thresholds for the same total concentration of Cu in the soil.

After leaching, it is also important that soils are aged to allow TEs to approach a partitioning (K_d) and lability (TE in rapid equilibrium with soil solution) similar to that of field soils. Aging will result in slow decreases in the amounts of labile TEs as they become incorporated into the solid phase (Barrow 1987; Bruemmer et al. 1988; Ma et al. 2006). There are many reactions that can reduce metal lability over time (McLaughlin 2001a), from solid-state diffusion (Barrow 1987; Ahmed et al. 2009) to occlusion in mineral micropores (Fischer et al. 1996) (Figure 2.12). From results published to date (Crout et al. 2006; Ma et al. 2006; Wendling et al. 2009), it is evident that aging is important to allow TEs to equilibrate with soil, and the period of aging used will be a compromise between the required speed of laboratory testing and the need to do testing in conditions similar to those in the field.

2.12 CORRECTION FACTORS FOR EXISTING TOXICITY DATA

Given the large number of toxicity studies conducted with spiked, unleached soils, there is a clear need for a means of translating from spiked laboratory soils to field soils. A number of studies have been carried out to find a "leaching factor" or a leaching-aging factor (LAF). In the study by Oorts et al. (2006b), they found that leaching

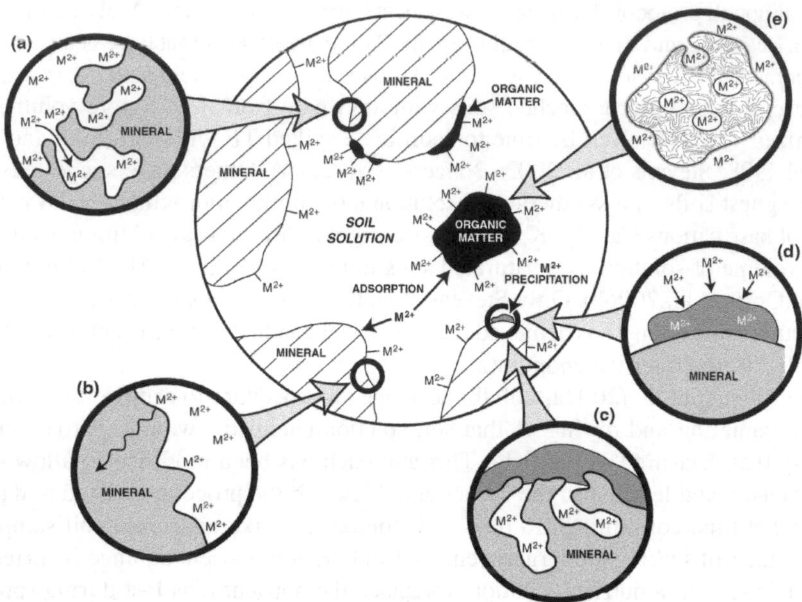

FIGURE 2.12 Aging reactions of TEs in soil – (a) micropore diffusion, (b) solid-state diffusion, (c) occlusion in micropores, (d) precipitation and Ostwald ripening, and (e) occlusion on organic matter. M^{2+} represents free divalent TE cations. Reprinted with permission from ICMM.

spiked soils reduced Cu toxicity by 1.3-fold, whereas a field leaching and aging treatment reduced toxicity by 2.3-fold. No significant differences were observed among aging treatments of between 6 and 18 months; thus they concluded that there was no aging effect based on time but that leaching accounts for much of the observed difference in toxicity between freshly spiked and aged soils. A more recent compilation and review of ecotoxicity testing for the European risk assessment process for Co, Cu, Ni, Pb, and Zn (Smolders et al. 2009) compared LAFs determined on a wide range of soils and for several endpoints (Figure 2.13). These LAFs were used to convert effect thresholds derived from laboratory ecotoxicity data to those used for field soils for the EU risk assessment process for these TEs. These LAFs have also been adopted into Flemish (Vlaams Reglement Bodemsanering [VLAREBO] 2008) and the proposed Australian SQSs for TEs (Warne et al. 2009).

2.13 RECOMMENDED "BEST PRACTICE" TE DOSING IN LABORATORY ECOTOXICITY EXPERIMENTS

The best practice is to spike soils under field conditions and allow natural processes of leaching and aging to proceed—this has been done in only a few experiments. Spiking to create soils with a range of trace metal concentrations should always be accompanied by a leaching step to reduce the ionic strength of the soil pore water to a value similar to that of the controls (e.g., McLaughlin et al. 2006; Broos et al. 2007a;

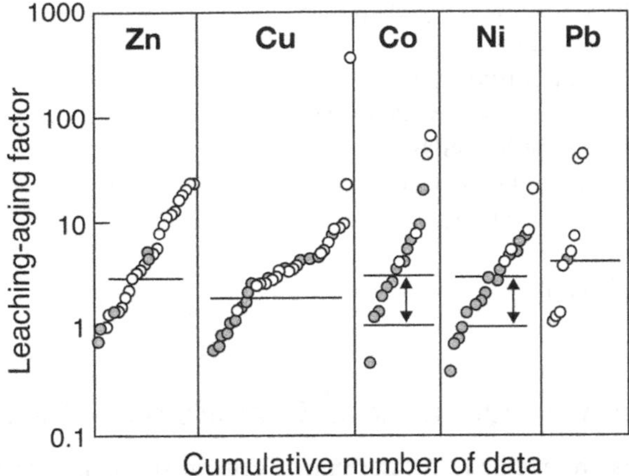

FIGURE 2.13 Leaching-aging factors for Co, Cu, Ni, Pb and Zn in soils determined using 10% effect doses (ED10 values) in freshly spiked soils and leached/aged soils. Closed symbols are bounded values, open symbols are unbounded values (i.e., minimum LAF values). Horizontal lines and ranges are the LAF factors selected for the EU risk assessment. For Co and Ni, a soil pH dependent relationship was used that is given here as a range (reproduced from Environmental Toxicology and Chemistry 28, Smolders et al., Toxicity of trace metals in soil as affected by soil type and aging after contamination: Using calibrated bioavailability models to set ecological soil standards, Copyright (2009), with permission from the Society of Environmental Toxicology and Chemistry).

Warne et al. 2008b). Because spiking can remove nutrient cations from the cation exchange sites, the leaching procedure should consist of a nutrient solution capable of reestablishing the balance between nutrient cations on the exchange sites consistent with optimum function of the organism or process being studied. In all cases, soil pore water should be collected and analyzed to verify that the conditions are similar to those in the field and so that the relationship of toxic response to soil pore water chemistry can be established. After leaching, soils should be aged prior to ecotoxicity testing, preferably for 30 days or more.

2.14 BIOTIC FACTORS AFFECTING ORGANISM RESPONSE TO TE DOSE

As noted in the Introduction to this chapter, there are several biological factors related to the ecotoxicological response by soil organisms to TEs in soils that can also markedly influence the development of robust SQSs. These are outlined below and treated in more detail in the subsequent sections:

Quantity and quality of the ecotoxicological data
Minimum number of ecotoxicity data points
Taxonomic diversity needed
Selection of species

Appropriateness of ecotoxicity endpoints
Type of ecotoxicity data
Use of acute and/or chronic data
Dealing with multiple toxicity data for species
Choice of distribution for SSDs
Level of protection to be provided
Acclimation and adaptation
Mixture considerations
Secondary poisoning

2.14.1 Quantity and Quality of the Ecotoxicological Data

Different options are available for deriving a critical soil concentration for use as an SQS. Two approaches have been widely used. The first is the application of an assessment factor (AF) to relevant ecotoxicity data. The AF approach is typically used for "data poor substances"; that is, substances for which few relevant and reliable ecotoxicity data are available. In this approach, the lowest toxicity value is divided by an AF (also called safety, adjustment, or uncertainty factor) to obtain the critical soil concentration. The magnitude of the AF that is applied is a function of the quantity and nature of the ecotoxicity data. The more ecologically relevant the data (e.g., chronic data are more relevant than acute data) and the greater the number of species and types of organisms for which there are data, the smaller the AF used. Typically, 3 AFs are used in this approach, each with a value of 10. The AFs are used to account for interspecies variation, differences between acute and chronic toxicity, and differences between laboratory and field toxicity data. These range from 1000 (if very limited numbers of acute EC50 values are available) to 10 (if reliable and relevant chronic ecotoxicity data are available for 3 trophic levels (e.g., European Commission [EC] 2003).

The main advantage of the AF approach is that, by focusing on the most sensitive available data value, it yields a protective value and the method is simple for users to understand. However, the choice of magnitude of the AFs is typically an arbitrary decision (Kooijman 1987; Okkerman et al. 1991; Organisation for Economic Co-operation Development [OECD] 1992; Schudoma 1994; OECD 1995; Rand et al. 1995; Chapman et al. 1996; Warne 1998). The approach also may be problematic as it does not consider the essentiality of many TEs to soil-dwelling organisms (e.g., B, Cu, Ni, Zn). Given the parabola-shaped concentration response curve for TEs (where adverse responses can be caused by either a deficit or an excess of TE), it is feasible that the application of a large AF could result in an SQS that is below the concentration required by organisms. In an attempt to address this, one of the default AFs of 10 used for TEs in the Australian and New Zealand water quality guidelines (WQGs) was reduced to 2 (Australian and New Zealand Environment and Conservation Council [ANZECC] and Agriculture and Resource Management Council of Australia and New Zealand [ARMCANZ] 2000). However, the issue of essentiality of TEs and how to address this in the derivation of SQSs for TEs has not been resolved (Chapter 4).

Although efforts have been made to quantify uncertainty (e.g., in terms of inter- and intraspecies variability) to support the magnitude of AFs (e.g., Chapman et al. 1996; Bonnell Environmental Consulting 1999), for the most part, there is little scientific basis to them (see review by Warne 1998). Additionally, by focusing on one ecotoxicity value, the AF approach ignores other relevant information and is not consistent with the risk-based paradigm. Given these limitations, the generally preferred method to derive SQSs is to use an SSD. This is the case in Austria, Belgium, Czech Republic, Denmark, Finland, France, Germany, Italy, Lithuania, Netherlands, Poland, Slovenia, Sweden, the United Kingdom, and for the proposed Australian methodology for deriving SQS (Heemsbergen et al. 2008). These SSD methods require more data than the assessment factor approach and are therefore only used for "data-rich substances."

When SSD methods were first proposed and used to develop environmental quality guidelines (EQGs), there was considerable discussion in the scientific literature about the merits and scientific validity of the approach and whether they were in fact any better than the existing AF methods (Calabrese and Baldwin 1993; Forbes and Forbes 1993; Smith and Cairns 1993; Schudoma 1994). However, a combination of factors such as the experimental validation work conducted by Okkerman et al. (1991, 1993), Emans et al. (1993), Hose and Van den Brink (2004), and Kefford et al. (2006); greater acceptance of the risk paradigm (see the seminal book on SSDs by Posthuma et al. 2001); and the advantages of these methods over the assessment factor method to those that implement the EQGs, has meant that they are now widely accepted. Most current water and SQSs now have a hierarchical system in which SSD methods are used in preference to the AF method to derive SQS, providing there are sufficient data available.

The goal of the SSD approach is to quantify the range of interspecies variability, and the outcome of this analysis is the concentration that corresponds to a predefined point on the cumulative distribution of ecotoxicity data (typically, the lower 5th percentile is used, which is generally referred to as the hazardous concentration to 5% of species (HC5) or the protective concentration for 95% of species (PC95). The variability surrounding any particular HC value is influenced by the distribution of the data and also the number and type of data used in the calculation. To reduce the influence of the data on the outputs from an SSD, minimum data requirements are needed in terms of both the number of species and the number of types of organisms (taxonomic groups) (see Sections 2.14.2 and 2.14.3).

The quality of the critical soil concentration generated by either an AF or SSD method is intimately linked to the quality of the toxicity data used in their derivation—as the maxim goes "garbage in equals garbage out." It is therefore crucial that all collated toxicity data have their quality assessed and that only data of acceptable quality are used to derive SQSs. While most SQSs state that the data that they have used have been assessed for quality, inevitably no information on this process is provided. Exceptions to this are the data assessment methods of Heemsbergen et al. (2008, 2009) (see Table 2.3) and MERAG (http://www.icmm.com/document/255) for evaluating the reliability and relevance of soil ecotoxicity data.

TABLE 2.3
Suggested scheme to assess the quality of terrestrial ecotoxicology data for trace elements[a]

Question	Score
Was the duration of the exposure stated?	10 or 0
Was the biological endpoint (e.g., immobilization or population growth) stated and defined (10 marks)? Was the biological endpoint only stated (5 marks).	10, 5 or 0
Was the biological effect stated (e.g., LC or NOEC)?	5 or 0
Was the biological effect quantified (e.g., 50% effect, 25% effect)? The effect size for NOEC and LOEC data must be quantified.	5 or 0
Were appropriate controls (e.g., a no-TE control and/or solvent control) used?	5 or 0
Was each control and TE concentration at least duplicated?	5 or 0
Were test acceptability criteria stated (e.g., mortality in controls must not exceed a certain percentage) (5 marks)? OR	5, 2 or 0
Were test acceptability criteria inferred (e.g., a standardised test method used which has validation criteria) (2 marks). Note: Invalid data must not be included in the database.	
Were the characteristics of the test organism (e.g., length, mass, age) stated?	5 or 0
Was the type of test media used stated?	5 or 0
Were the TE concentrations measured?	4 or 0
Were parallel reference TE toxicity tests conducted?	4 or 0
Was there a concentration-response relationship either observable or stated?	4 or 0
Was an appropriate statistical method used to determine the toxicity?	4 or 0
For NOEC/LOEC data was the significance level 0.05 or less? OR	
For LC/EC data was an estimate of variability provided?	4 or 0
Were the following parameters measured and stated (3 marks each) or simply stated (1 mark each)? pH, organic matter or organic carbon content, clay content, cation exchange capacity.	12, 4 or 0
Was the temperature measured and stated?	3 or 0
Was the grade or purity of the test TE stated?	3 or 0
Were other cations and or major soil elements measured? OR	
Were known interacting elements on bioavailability measured (e.g., Mo for Cu and Cl for Cd)?	3 or 0
For spiked soils with TE salts: Were the soils leached after spiking?	3 or 0
Were the incubation conditions and duration stated? (both 3 marks, only one 1 mark).	3, 1 or 0
Total score	
Total possible score for the various types of data and contaminants.	102
Quality score ([Total score/102] × 100)	
Quality class (H ≥80%, 51–79% A, U ≤ 50%)[b]	

[a] Table reproduced from Heemsbergen et al. 2008, 2009. Copyright (2009), with permission from Elsevier.

[b] H, high quality; A, acceptable quality; U, unacceptable quality.

2.14.2 Minimum Number of Ecotoxicity Data Points

For data-poor substances, the AF method focuses attention on the lowest threshold value obtained, and, as outlined above, the magnitude of the AF is a function of the number, type, and relevance of toxicity data available. The minimum number of data required to use this method is 1.

For statistical methods such as the SSD, the more data that are used the more powerful and reliable are the results of the analysis. The Danish EPA (Pedersen et al. 1994) and the OECD (1995) found that EQGs derived using the SSD approach with data for less than 5 species were highly dependent on the spread of the data, while for EQGs derived using data for 5 or more species this effect was markedly reduced. These findings were widely adopted as the minimum data requirement to use SSD methods in deriving EQGs (e.g., OECD 1992, 1995; Van de Plassche et al. 1993; ANZECC and ARMCANZ 2000; Crommentuijn et al. 2000b). Newman et al. (2000) used nonparametric methods to estimate that the optimal number of species needed to minimize variation in EQG variation due to the random replacement of values ranged from 15 to 55 (with a median of 30). Wheeler et al. (2002) suggested that a minimum of 10 to 15 species were needed. Reflecting these findings the EU has recommended (EC 2003; ECHA 2008) that the minimum data requirement to use an SSD for aquatic ecosystems is toxicity data for 10 species that belong to 8 taxonomic groups. The minimum data requirement to use an SSD method is ultimately a pragmatic compromise between the optimal number and the number of TEs for which the minimum data requirement will be met. Thus, from a scientific rigor point of view the minimum data requirement should be at least 10 species. However, in reality this would mean that EQGs could only be developed using SSD methods for a limited number of contaminants, thus maximizing the number of contaminants that would have EQGs derived by the AF method. Reflecting the general paucity of toxicity tests for soil organisms and a strong preference to use the SSD approach, the proposed methodology for deriving Australian soil quality guidelines (Heemsbergen et al. 2008) has a minimum toxicity data requirement for using an SSD approach of toxicity data from 5 species that belong to 3 taxonomic groups, while the Dutch (Crommentuijn et al. 2000a) and most EU member countries require toxicity data for at least 4 species (Carlon 2007).

2.14.3 Taxonomic Diversity Needed

As noted above, for data-poor substances where the AF method is used, there is no minimum taxonomic diversity in ecotoxicity endpoints needed to derive an SQS. Less diverse data simply result in a higher AF.

There is very limited guidance available on the minimum amount of taxonomic diversity needed in the data to derive SQSs by an SSD method. Considerably more guidance is available for WQGs, and as the issues regarding this are essentially the same and there is more information for aquatic ecosystems, this will form the basis of the following discussion. The requirements of various jurisdictions in terms of the number of organism types are not consistent. Both the United States (Stephan et al. 1985) and the EU (EC 2003) require toxicity data for species that belong to 8 different taxonomic groups. Current WQGs for most European countries, Australia and New Zealand, Hong Kong, and South Africa require toxicity data for species that belong to at least 4 different taxonomic groups. As with the decision regarding the total number of species for which toxicity data are required, the decision regarding the required degree of biodiversity in the toxicity data is a compromise between scientific rigor and pragmatism. Given the lack of soil toxicity tests, it would not be sensible to require that the same degree of biodiversity in the toxicity data for soil ecosystems as for aquatic ecosystems.

The requirements of the EU and USEPA methods for toxicity data to belong to specific organism types, in fact, invalidate one of the key assumptions of all SSD methods—that the species that are used are a random selection from those in the ecosystem being protected or assessed (Calabrese and Baldwin 1993; Forbes and Forbes 1993; Smith and Cairns 1993; Schudoma 1994). Therefore it is argued that the requirements for biodiversity in the toxicity data should not specify the taxonomic type of organism but rather simply require that a minimum number of different taxonomic groups are represented as is the case in the proposed Australian methodology for deriving SQSs (Heemsbergen et al. 2008).

There are currently standardized soil toxicity tests for plants, soil invertebrates, and microbial processes. Therefore it is recommended that toxicity data are required for at least 3 taxonomic groups to derive SQSs using an SSD method. However, it is possible to identify unique taxonomic groups within these 3 taxonomic groups. For example, plants could be divided into monocots and dicots, invertebrates could be divided into soft bodied (e.g., earthworm) and hard bodied (e.g., colembola) organisms, and microbial processes could be divided into the type of biochemical reaction that is being quantified (e.g., respiration, nitrification, and denitrification). Indeed this approach, apart from dividing plants into 2 groups, has been adopted in proposed Australian SQS methodology (Heemsbergen et al. 2008). Thus, as more standardized soil toxicity tests are derived and are used to generate toxicity data, it is recommended that the minimum number of taxonomic groups required be increased.

2.14.4 Selection of Species

There is a divergence of opinion and practice regarding whether SQSs should be derived using all vertebrate, plants, and soil organism toxicity data or whether more restrictive approaches should be used. If the goal is a more refined protection target (e.g., for a particular land use or geographic or climatic region), then ideally only the ecologically relevant species for that land use or region should be selected. An example of how species might be selected to suit a site with particular soil physicochemical properties is shown Figure 2.14 (Jansch et al. 2005).

For instance, despite the common use of *E. albidus* in ecotoxicity testing, several test species of enchytraeids could also be considered depending on their edaphic preferences. For example, if the jurisdiction includes more acidic soils, *E. crypticus* would be the best option. Other groups are fairly broad concerning their ecological requirements, such as *F. candida*, and would fit most scenarios. Additionally, one can also consider using data for endemic species if available.

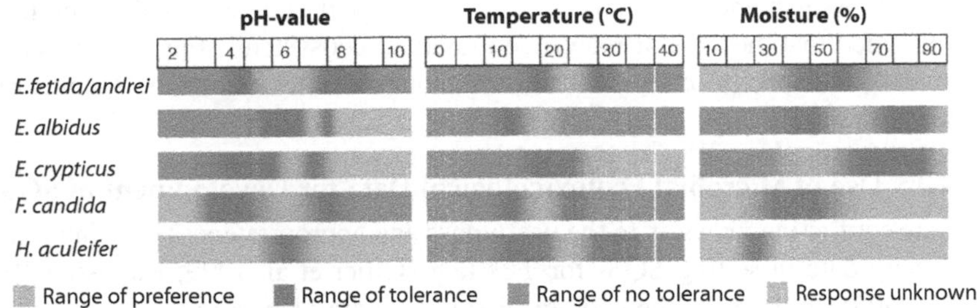

FIGURE 2.14 Ecological ranges of common terrestrial ecotoxicological test species regarding pH-value, temperature and moisture (reproduced from Environmental Reviews 13, Jänsch et al., Identification of the ecological requirements of important terrestrial ecotoxicological test species, Copyright (2005), with permission from NRC Research Press).

Use of aquatic species toxicity data for terrestrial SQSs is probably inappropriate, as the species present and the physicochemical environment are extremely different; e.g., soils have a much wider range of pH than natural waters.

As stated in the previous section, it is not appropriate to mandate the use of toxicity data for particular types of organisms. It is equally inappropriate to require the use of specific species. Ideally, the species used should be a random selection of species from, or relevant to, the ecosystem for which the SQSs are being derived. For example, developing SQSs in Asia should take into account the extensive use of flooded land for rice cultivation. Such considerations may demand the development of new toxicity test methods and a thorough characterization of the behavior of TEs in such systems.

Because of the general paucity of toxicity data for soil organisms, it is not always possible to only use relevant native species, and in many cases SQSs will have to be derived entirely using toxicity data for nonnative species (e.g., some SQSs for Australia being derived exclusively using toxicity data for North American and European species). By using nonnative species the assumption is made that they have the same range of sensitivities as the native species. The validity of this assumption has been questioned and examined in a number of studies using aquatic species (e.g., Dyer et al. 1997; Markich and Camilleri 1997; Brix et al. 2001; Hobbs et al. 2004; Hose and Van den Brink 2004; Maltby et al. 2005; Chapman et al. 2006; Kwok et al. 2007). However, the evidence is conflicting with some studies (Maltby et al. 2005; Hose and Van den Brink 2004) finding no differences, while others have found chemical-specific differences (Dyer et al. 1997; Markich and Camilleri 1997; Brix et al. 2001; Hobbs et al. 2004; Chapman et al. 2006; Kwok et al. 2007). This led Chapman et al. (2006) to conclude that "toxicity data from one geographic region will not be universally protective of other regions" (p. 1081).

The fact that soil chemistry can vary markedly from one geographic region to another and that soil physicochemical properties can modify toxicity (see earlier sections) may exacerbate or ameliorate differences in the sensitivity of organisms from different regions. It is therefore recommended that until there are sufficient toxicity data for native species to meet the minimum data requirements of using the SSD approach, that native and nonnative data are pooled and used to derive SQSs. For generic risk assessments, as many diverse species and taxa as possible should be used, provided that they are of relevance to natural ecosystems. For higher tier risk assessments, the species or taxa considered should be relevant for the specific ecosystem or land use.

2.14.4.1 Use of Microbial Ecotoxicological Data for Development of SQS

There are conflicting views as to the usefulness and appropriateness of using microbial toxicity data in setting SQSs for TEs (e.g., Giller et al. 1999; Kapustka 1999; USEPA 2005). In deriving the ecological soil screening levels in the United States (Eco-SSLs), the USEPA stated that "Like amphibians and reptiles, EPA recognizes their importance within terrestrial systems, but concurs with the workgroup that data are insufficient and the interpretation of test results too uncertain for establishing risk-based thresholds" (USEPA 2005, p. ES-2). However, in the EU, microbial toxicity data form a central part of the development of SQSs, and Giller et al. (2009) recently argued that microbial endpoints should be considered in the development of SQSs as the science has advanced greatly over the last 5 years owing to the research undertaken for the EU risk assessment process.

To place the techniques in perspective, it is worth considering the extent to which techniques have been adopted and then relating them to the criteria required for useful and appropriate soil ecotoxicity assays, mainly ease of use, responsiveness, robustness, and reliability.

Soil microbial carbon. The soil microbial biomass is defined as organisms living in soil that are generally smaller than approximately 10 μm. Most attention is given to fungi and bacteria because they generally dominate the biomass, but any measure includes both active and dormant microorganisms. Jenkinson et al. (1987) described biomass carbon as "the eye of the needle through which all organic matter needs to pass," and as such it is widely and effectively used in studies of TEs (Fritze et al. 1996). It has been suggested that the microbial biomass content is an integrative signal of the microbial significance in soils because it is one of the few fractions of soil organic matter that is biologically meaningful, sensitive to pollution, and demonstrably measurable. The widespread adoption of 4 indirect methods: fumigation-incubation, SIR, fumigation-extraction, and ATP content (Jenkinson and Powlson 1976; Anderson and Domsch 1978; Jenkinson and Ladd 1981; Vance et al. 1987) has revolutionized soil biological characterization. However, it has been demonstrated that biomass measurements are noisy and that the variability occurs with little direct correlation to soil quality (Martens 1995; Dilly and Kutsch 2000; Broos et al. 2007b).

Microbial diversity structure. Originally, diversity of microorganisms in soil was studied using the culture plate method, where organisms were plated onto various media and growth observed visually. Although the approach is relatively simple, the data are often unclear and endpoints difficult to interpret. Bundy et al. (2004) reported that the method was insensitive to treatment factors but prone to responding to incubation and preparatory stages. Ritz (2007) proposed that such techniques were of little value because of the small percentage of the microbial population able to be cultured and that data were therefore highly biased.

More sophisticated methods such as the assessment of phospholipid fatty acid and fatty acid methyl ester profiles were developed decades ago and continue to be widely used. These techniques rely on extraction and identification of groups of cell compounds specific for microbial groups in the soil. Such quantification offers a much higher resolution of the biomass extractable fraction and standard multivariate approaches have been used to show these techniques to be useful for examining TE-contaminated soils (Kirk et al. 2004). The approach enables categorical identification of key functional groups in soils and is appropriate both for fungal and bacterial communities. Ease of access to gas chromatography-mass spectrometry (GC-MS) and relative procedural automation should make this technique (which requires no culturing steps) more widely used. However, the perception of time-consuming steps and high sensitivity combined with great confusion in data analysis has meant that these approaches remain as research tools rather than routinely adopted procedures.

Molecular techniques have demonstrated that little is known about the vast majority of the organisms that constitute the microbial biomass. Direct methods, independent from cultivation, based on the genotype (Amann et al. 1995) and phenotype (Zelles 1996) of the microbial composition have allowed an insight into the limitations of traditional approaches. Using, for example, the rDNA directed approach of dissecting bacterial communities by amplifying the 16S rDNA gene from samples by polymerase chain reaction (PCR) and studying the resultant sequences, it has become apparent that new sequences with little relationship to well-characterized isolates are prevalent (Amann et al. 1995). Furthermore, frequently occurring, yet uncultured bacteria became visible microscopically through the application of fluorescently labeled rRNA-directed oligonucleotide probes. It is a consequence of these technologies that researchers have estimated that 1 g of soil consists of more than 10^9 bacteria belonging to as many as 10,000 different microbial species (Ovreas and Torsvik 1998).

The complexity of this diversity calculation makes it almost impossible to use as a meaningful indictor of changes in soil. There are a limited number of studies that have demonstrated clear effects of TE doses on microbial diversity. Muller et al. (2001) investigated the effects of mercury on the soil microbial community and reported that bacterial diversity decreased with dose, but there was no difference in fungal biomass. This is a widely reported observation. Kirk et al. (2004) critically appraised the availability of molecular techniques in evaluating the soil community and revealed that many methods may have constraints if applied to TE studies. Ambiguous endpoints and biased sampling steps were reported to be the greatest limitations.

Recently, advances in genomic analysis and stable isotope probing (SIP) have allowed a categorical link to be made between structure and function in microbial communities. This link has been made by the combination of traditional biochemical process measurements with modern molecular strategies. These techniques are, however, expensive. SIPs require the use and detection of ^{13}C, which may cause bias because of the way in which substrates are introduced. Metagenomics and proteonimics are now being applied to soils but are still far from regulatory application.

Another approach is to use molecular techniques to study indicator organisms, for example, beneficial symbionts such as *Rhizobium* or arbuscular mycorrhizae. Rhizobia populations are essential to increase the yield of leguminous crops. Although little is known about the evolution of natural bacterial populations in relation to land management, rhizobia diversity and abundance is sensitive to TE amendments. Despite the widely reported sensitivity of *R. leguminosarum* biovar *trifolii* to TE using traditional culture assays, there is an absence of supportive molecular evidence. Indeed, most molecular probes have been designed for other legumes.

Microbial activity. Soil microbial activity leads to the release and cycling of nutrients in the pedosphere and is an essential component of the biogeochemical cycle. Microbial activity is regulated by a wide range of soil physical and chemical parameters. The wide selection of techniques used to measure components of microbial activity allows a relative quantification of the metabolic processes of the given communities. Microbial activity measurements can be made either during long-term field measurements or short-term laboratory experiments.

Experiments in the field require lengthy periods of incubation (Hatch et al. 1991; Alves et al. 1993) before significant changes of product concentrations are detected. Four to 8 weeks was reported by Hatch et al. (1991) and Alves et al. (1993) for an estimation of net N mineralization. Field measurements are often difficult to interpret. For example, field soil respiration values do not exclusively represent microbial components and the respiration associated with plants will vary greatly throughout diurnal and seasonal periods (Dilly and Kutsch 2000). Short-term laboratory assays conducted with sieved samples under standardized conditions offer greater reproducibility and specificity. This is why laboratory-based respiration using prepared samples is so widely adopted.

Microbial activity measurements also include enzymatic assays making use of measuring substrate-specific transformations (Burns 1977). Laboratory methods can be standardized for comparative evaluation of contrasting soils under varied treatments in different laboratories. Taylor et al. (2002) proposed 2 primary reasons for measuring soil enzymes. First, measurements represent process diversity hence an evaluation of the biochemical potential and the possible resilience of the soil can be made. Second, enzymes report on key functions and activities that can provide a temporal measure of the impact of a given TE dose. Furthermore, enzymatic measurements may represent the redundancy of the soil biochemical system and thus provide a potential value of resilience.

Despite these great potential advantages, Pettit et al. (1977) identified limitations such as the fact that the values reflect a measure of the potential activity, which is encoded in the "soil genotype": but in real environmental conditions this is rarely achieved. Enzymatic assays enable an evaluation of "soil functional groups" but they are conducted under optimal environmental conditions with no substrate limitations. Furthermore, optimizing conditions for the enzyme assay (e.g., changing media pH) can change the speciation and bioavailability of TEs compared to the conditions naturally present in the soil. There are at least 500 enzymes with critical roles in the cycling of C or N or both, and until recently such measurements could not be made routinely (He et al. 2007). If, on the one hand, there is genuine redundancy in enzymatic functions in soil, the loss of a specific enzyme or several enzymes should not have a major effect. If, on the other hand, changes in the activity of some "benchmark" enzymes provide an early indication of changes in process diversity, soil enzymatic measurements have a clear role in assessing TE impacts.

Nitrogen turnover assays. Bioavailable nitrogen is essential for plant growth; hence nitrogen processing is a key component of a functioning soil. Two main delivery processes (mineralization and nitrogen fixation) are related to nitrification and denitrification rates. Measurements of these fluxes are key to understanding the given performance of a soil.

The microbial mineralization of proteins in soils mobilizes organically bound nitrogen (Ladd and Butler 1972), and it is widely acknowledged that there is low substrate specificity (Kalisz 1988). Conversion of ammonium to nitrate is performed by a relatively narrow range of organisms (bacteria and archaea) (Mertens et al. 2010) and is regarded as a sensitive toxicity endpoint affected by a range of TEs (Smolders et al. 2001).

Nitrogen fixation is acknowledged to be relatively sensitive to TE additions to soil (Giller et al. 2009), and most work has focused on symbiotic organisms such as *Rhizobium*, which has been found to be relatively sensitive to TE additions (Chaudri et al. 1992). Free-living diazotrophs have received little attention because they are associated with low fixation rates, yet they may be the most sensitive indicator of change. Molecular approaches are based on PCR amplification of the nifH gene and its mRNA transcripts for the group-specific detection of free-living diazotrophs in soil (Widmer et al. 1999).

Adoption of microbial assays in soil ecotoxicity assessments and their comparative application. While process-level microbial methods with functional redundancy (e.g., basal respiration or SIR) are simple to perform, they continue to be acknowledged as being relatively insensitive to TE additions. Tests that assess the performance of defined populations of a few species with key functions, e.g., nitrogen fixation, nitrification, tend to be more sensitive. However, there appears to be a trade-off between sensitivity of a microbial ecotoxicity test and its robustness (variability in uncontaminated soils). Broos et al. (2005) compared a range of microbial and plant ecotoxicity tests across 7 series of field-contaminated soils and found an inverse relationship between robustness and sensitivity (Figure 2.15). The most commonly used and robust sentinel assays (for comparative purposes) involve the use of earthworms or springtails, because their relevance is based on the premise that selected representative species that graze on microbial or plant substrates will be TE sensitive. However, in EU risk assessment research programs, soil invertebrates have generally been found to be insensitive endpoints for TE contamination of soils. Hence a focus on the most sensitive endpoints may be counterproductive, as the assay also needs to be robust across a range of uncontaminated soils. Adverse effects of TE contamination to an endpoint that is not robust are therefore unlikely to have large ecological effects, as the endpoint is already variable in natural soils. For example, measurement of microbial biomass C in soil is often used to assess the effects of TE on the soil biota, but Broos et al. (2007b) demonstrated that in a single uncontaminated field, this parameter varied 4-fold, thus requiring unreasonably large numbers of replicates (9 to 93) to permit the determination of a 20% change in microbial biomass C by hypothesis-based statistical techniques at $p \leq 0.05$.

Despite these problems with some microbial assays, the importance of the microbial biomass in soil and the functions that it performs requires that microbial ecotoxicity data be included in the development of SQS. At the fundamental level, soil microbiology continues to present challenges as to how researchers should link the presence of individual bacteria, fungi, and other components to the key soil processes:

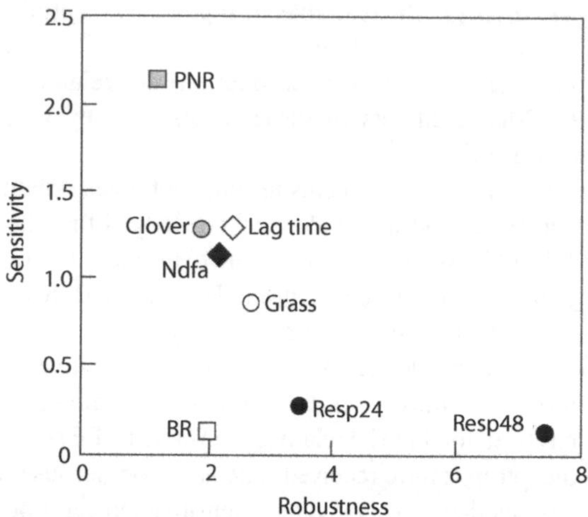

FIGURE 2.15 Robustness (mean/standard deviation in uncontaminated soils) versus TE sensitivity (arbitrary scale based on LOEC data) scores for different ecotoxicological endpoints. N_{dfa} = percentage N fixed by clover; Clover = clover yield; Grass = grass yield; PNR = potential nitrification rate; Lag = lag time for substrate-induced respiration; Resp24 and Resp48 = cumulative respiration 24 and 48 hours after substrate addition, respectively; BR = basal respiration (reproduced from Environmental Toxicology and Chemistry 24, Broos et al., Toxicity of heavy metals in soil assessed with various soil microbial and plant growth assays: a comparative study, Copyright (2005), with permission from the Society of Environmental Toxicology and Chemistry).

substrate metabolism, performance in food webs, and role in nutrient cycling in relation to TE contamination of soils. Although it is possible to measure molecular and phenotypic diversity, as well as the more usual biochemical processes and functions in soil, drawing these 2 disciplines together when dealing with the impact of TEs in soil has been ineffective to date. However, current and future developments in this area are likely to lead to greater regulatory acceptance of microbial endpoints when assessing TE toxicity in soils (Giller et al. 2009).

2.14.5 Appropriateness of Toxicity Endpoints

An aim of an SQS may be to protect organisms that comprise ecosystems from chronic sublethal effects that arise due to lifelong exposure and from acute lethal effects that arise from short-term exposure. Depending on the aim of the SQS, various toxicological endpoint data may or may not be appropriate. The currently available toxicological endpoints range from subcellular through organismal to population, community and ecosystem level. It is generally considered that endpoints below the individual or organismal level are only indicators of exposure to contaminants rather than measurements of the effects that arise from the exposure and are therefore inappropriate to use in generating SQSs (Figure 2.16).

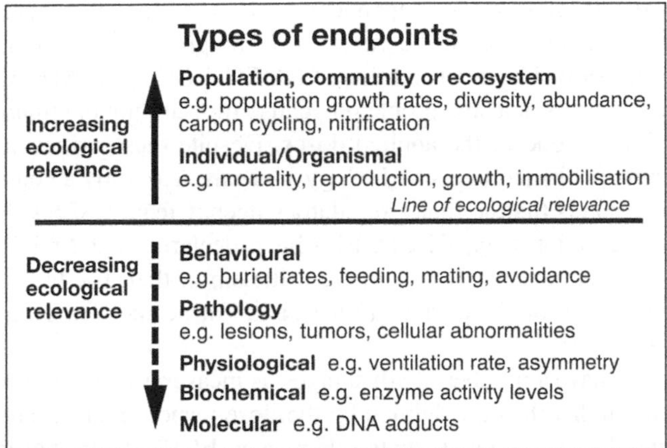

FIGURE 2.16 The types of measurable toxicological endpoints and the generally accepted point at which the ecologically relevant and low relevance endpoints are separated.

The ideal toxicity data to derive SQSs are field-based values that measure ecosystem level effects. However, since these data are not readily available, toxicity data with lower levels of environmental relevance are used in the derivation of SQSs. Nonetheless, the toxic endpoints used should have the most direct theoretical and practical link with population or higher-level effects, i.e., survival, reproduction, population growth rate, and possibly growth. The availability of such toxicity data is very much organism dependent—there is little available for plants and environmentally relevant vertebrates but usually relatively large amounts for invertebrates and microbes.

Considerable effort is being expended in trying to provide evidence that endpoints at the suborganismal level have ecological relevance; this may be provided in the future. However, until this occurs, only endpoints that measure effects at the individual or higher levels of organization should be used to derive SQSs.

2.14.6 Type of Ecotoxicity Data

Toxicity data are either point estimate measures or hypothesis-based measures. The former consists of lethal (LC), effect (EC), or inhibition (IC) concentration data that denote a certain percentages of the test population or the endpoint being measured (e.g., enzyme activity) is affected by 50, 20, and 10%. The latter consist of NOEC, lowest observed effect concentration (LOEC), and maximum acceptable threshold concentration (MATC, which is the geometric mean of an NOEC and an LOEC).

Hypothesis-based toxicity data (i.e., NOEC, LOEC, and MATC) have been criticized and said to be inappropriate for use in deriving EQGs since the early 1990s (e.g., Hoekstra and Van Ewijk 1993; Noppert et al. 1994; Chapman et al. 1996). Issues associated with the use of hypothesis-based toxicity data have been clearly defined: the value is arbitrary and dependent on the selection of concentrations used in the bioassay and is subject to the vagaries of the precision of chemical analysis (Chapman et al. 1996). The incorporation of NOEC, LOEC, and MATC into subsequent statistical models is also a problem because they are not continuous variables but by definition are one of the applied doses. Despite these criticisms, such data are still routinely published in ecotoxicology journals. Certainly a contributing factor for this situation is that no major regulatory agency (e.g., USEPA, Environment Canada) or organization (e.g., OECD, EU) have declared that NOEC and LOEC type data should not be used. However, the criticism of their use has recently been reinvigorated by Fox (2008), Newman (2008), and Warne and Van Dam (2008) with strongly argued papers.

ECx-type data have a number of advantages as measures of toxicity over hypothesis-based data, in that they are defined by the dose response curve, and from these 95% confidence limits can be estimated. However, ECx data are not without their problems. A sufficient number of concentrations are needed to precisely establish a dose-response curve from which ECx values can be estimated. Even then, unacceptable variability around an EC10 value may limit its application in SQS derivation. Recently, the EU has supported the use of calculated EC10 values in lieu of NOECs, with the general supporting argument being that a 10% effect is within the range of control performance. However, emphasis on the use of EC10 values in lieu of NOEC values has to be balanced with the difficulty in accurately measuring EC10 values owing to the high variability that can occur at this low level of response.

The majority of chronic toxicity data are NOEC, LOEC, and MATC type data, while the majority of acute toxicity data are point estimates. Even if hypothesis-based data were no longer produced from today onward, it will be necessary for some time into the future to continue using these data to derive SQSs. It is recommended that NOEC, LOEC, and MATC data continue to be used in combination with equivalent point estimate data to derive SQSs until there is sufficient point estimate data to meet the minimum data requirements of the SSD approach, at which point NOEC, LOEC, and MATC data should no longer be used.

2.14.7 Use of Acute and Chronic Data

There is little agreement over the definitions of acute and chronic toxicity data—they are often operationally defined. Acute toxicity data are those that are based on the short-term exposure (typically of not more than 96-hour duration) of test organisms/systems to a toxicant while chronic data are those based on longer-term exposure (longer than 96 hours or at least 10% of the lifespan of the organism). However, the life span of the test organism needs to be considered, and for this reason toxicity tests of 72 hours duration for micro-organisms are sometimes considered chronic. While chronic effect data are preferred for the derivation of SQSs, lack of chronic data often results in small data sets for the derivation, resulting in wide confidence intervals associated with SQS estimates. Suggestions have been made that acute data may be used to supplement the data set used in SQS derivation through the application of acute-chronic ratios (ACRs). Although the ability to increase the size of the data set by this means is appealing, it is not without problems. Acute modes of toxic action may differ from chronic modes of toxic action for the same chemical (including TEs) so a simple toxicity ratio may not be mechanistically supported. ACRs will also vary for different chemicals so sufficient data must be available for their estimation. The pooling of acute and chronic data should not be done unless there is solid evidence that acute and chronic effects can be related to a common mode of action. However, in the case where well-defined ACRs exist and limited chronic data are available, or where SQSs may be developed for assessing acute effects (e.g., chemical spills), acute data may be useful in SQS derivation.

2.14.8 Dealing with Multiple Toxicity Data for Species

For many TEs there are multiple toxicity data for a given soil organism. If all the data are used in an SSD, then the resulting output would not protect a selected percentage of species but rather a selected percentage of the toxicity data. There are several options available in such situations that ensure the resulting SQS protects a selected percentage of species.

The simplest way to handle multiple toxicity data for species is to use the lowest toxicity value for each species. This is a conservative approach, which may generate a low SQS and provide a level of comfort to the assessor, especially when a wide range occurs between the lowest and highest data points for a given species.

The most commonly used method to handle multiple toxicity data for species is to use the geometric mean (i.e., the normal method of determining the mean is applied to logged data and then the resulting value is antilogged). The geometric mean is used because it reduces the effect of extremely low or high values. If multiple toxicity data from a species are available for one endpoint (e.g., lethality), then the geometric mean is determined and used in the SSD to represent that species. If there are multiple data for multiple endpoints (e.g., lethality, reproduction, and growth) to a species, then the geometric mean for each endpoint is determined and the lowest geometric mean is used to represent that species. This aggregation and calculation of geometric means should also consider whether the toxicity tests were conducted under similar physical and geochemical test conditions that should result in similar bioavailability for the TE (EC 2003). Ideally, toxicity data from multiple tests should be normalized via bioavailability models (refer to Sections 2.5 to 2.9 inclusive) to a common set of abiotic factors prior to the calculation of the geometric mean. This will reduce the influence of test conditions on the variability of test results and will provide a more objective view of intraspecies variability.

The SSD is an assessment of interspecies variability. Using the geometric mean of species toxicity data provides a better estimate of the sensitivity of the species than the lowest toxicity value approach, as it is based on all the toxicity data. Despite this, valuable information regarding intraspecies variability is still lost using the geometric mean. This is important because this information can be helpful to understand the biological processes that affect the sensitivity of species towards TE toxicity. It is therefore recommended that intrinsic information on effects of specific TEs, e.g., mechanisms of toxicity and factors affecting bioavailability, be taken into account when aggregating data from multiple tests.

A final approach uses all the toxicity data points available, but each data point is weighted so that each species has the same weight within the SSD. This approach incorporates the known intraspecies variation while ensuring that no species is given more importance than any other. Such a weighting approach has been used by Hickey et al. (2008). An alternate weighting approach suggested by Forbes and Calow (2002) and that has been implemented by Hose and Van den Brink (2004) is to weight the species to reflect the relative abundance of taxa in the ecosystem being protected.

It is recommended that weighting of species toxicity data be performed where possible, rather than data reduction. Most of the available software packages for SSD development do not have a weighting function for data input, and until this option is offered, data reduction using geometric means will likely continue.

2.14.9 Choice of Distribution for SSD

A range of SSD methods is available to calculate SQSs. These can be divided into those that apply a single distribution to the toxicity data, those that apply a range of distributions to the data, and nonparametric approaches. The single distribution methods include the log-triangular distribution (Stephan et al. 1985), the lognormal distribution (Wagner and Løkke 1991; Aldenberg and Jaworska 2000), and the log-logistic distribution (Aldenberg and Slob 1993).

The multiple distribution SSD methods and nonparametric SSD methods were developed because in many instances the single distribution SSD methods did not fit the toxicity data well (Kwok et al. 2007). For example, when deriving the draft Australian and New Zealand WQGs it was found that the log-logistic distribution did not fit the available toxicity data for approximately 1/3 of the chemicals for which WQGs were being derived (Warne, personal communication). Other reasons include the fact that there is no theoretical basis for assuming that the sensitivity of species will have any particular distribution; e.g., the log-logistic distribution was selected by Aldenburg and Slob because of mathematical simplicity rather than any biological basis. Multiple distribution SSD methods were first developed by Shao (2000), who used the Burr type III family of distributions to model toxicity data. This approach is incorporated into the BurrliOZ software (Campbell et al. 2000), which is publicly available (http://www.cmis.csiro.au/envir/burrlioz/) and is used in Australia and New Zealand (ANZECC and ARMCANZ 2000) Hong Kong, and South Africa for WQGs and the proposed Australian SQGs (Heemsbergen et al. 2008; Warne et al. 2009). Subsequently, others have also adopted a multiple distribution approach by testing the fit of a range of statistical distributions to the toxicity data and using the best fitting model to determine the EQG (e.g., lognormal, log-logistic, bootstrap, and bootstrap regression based on the log-logistic distributions used by Wheeler et al. 2002; the normal, logistic, exponential, triangular, uniform, Weibull, Pareto, gamma, beta, and extreme value distributions used by Maltby et al. 2002; and the lognormal, log-logistic, bootstrap, and bootstrap regression based on the log-logistic distribution and log-triangular distributions used by Kwok et al. 2007). Using the lognormal distribution as the default has been advocated in the EU as this would provide consistency and comparability among different TEs (EC 2003). However, as stated above, there is no theoretical basis for the use of the lognormal distribution or any other specific statistical distribution. In general, parametric distribution-based SSD methods should only be used if they are supported by Goodness of Fit testing. Another option would be to take the average of all frequency distributions that are supported by Goodness of Fit tests. This was the approach taken for the marine effects of Ni in the EU Existing Substances Risk Assessment (Denmark 2008).

Two nonparametric SSD methods have been developed including the basic bootstrap method (Newman et al. 2000) and the bootstrap regression approach (Grist et al. 2002). While these methods can overcome the fit problems of the distribution-based SSD methods, they typically require more data. For example, in order to determine the concentration that should only permit 10 and 20% of species to experience toxic effects, a minimum of 10 and 20 toxicity data points are required, respectively.

It is recommended that a multiple distribution SSD approach be adopted to derive SQSs for TEs within reason. In light of the previous discussion, overfitting needs to be avoided. If, however, no distribution fits the data adequately, it is recommended that nonparametric approaches be evaluated.

2.14.10 Level of Protection to Be Provided

One great advantage of the SSD approaches over the assessment factor approaches is that the level of protection provided by the SQSs can be varied to any selected percentage of species. This has been taken advantage of to derive a standard level of protection (i.e., the concentration that should protect 50% of species) and a longer-term aspirational level of protection (i.e., based on the concentration that should protect 95% of species) as in the Dutch SQSs (e.g., Van de Plassche et al. 1993; Ministry of Housing, Spatial Planning and Environment 2000) or land use–based SQSs, which provide different levels of protection as in the Canadian Council for Ministers for the Environment (CCME) (1996, 2006) and proposed Australian SQSs (Heemsbergen et al. 2008; Warne et al. 2009). For example, in Australia and Canada SQSs, agricultural and residential and parkland land are set to protect 80 and 75% of species and soil processes, respectively, whereas for commercial and industry 60 and 50% of the species and soil processes, respectively, are protected (Heemsbergen et al. 2008; CCME 1996, 2006).

The most frequently used level of protection in EQGs including SQSs is the concentration that should protect 95% of species (PC95) and conversely only permit 5% of species to experience toxic effects (HC5). However, while this is the most frequently used, there is no scientific basis to it. The first to use the PC95 was Stephan et al. (1985), who stated that the rationale for choosing to protect 95% of species was that "other fractions resulted in criteria that seemed too high or too low in comparison with the sets of date from which they were calculated" (p. 2). Similarly, there is no scientific basis behind the other levels of protection used in SQSs; rather, it is a policy decision that may consider other nonscientific factors such as economic and social considerations. No particular level of protection is recommended as this will depend on the legislative framework and key issues in each jurisdiction.

2.14.11 Acclimation and Adaptation

Because some TEs are essential to soil organisms, the concentration of essential TEs in the media used to culture or acclimatize the test organisms should be evaluated. This issue will be addressed elsewhere, but in general, toxicity data to be included in development of an SQS should be from toxicity tests in which the test organisms were cultured in concentrations that are clearly not deficient.

Another issue to consider is adaptation of organisms under field exposure to TEs, especially where indigenous biota or functions are being used to assess TE toxicity from the same soils. It is well known that organisms can adapt to elevated TE concentrations in soil (either geogenic or anthropogenic sources) (Witter et al. 2000; McLaughlin and Smolders, 2001; Rusk et al. 2004; Mertens et al. 2006; Giller et al. 2009), and indeed this has been used to identify soil contamination by TEs through pollution-induced community tolerance (PICT) (Rutgers et al. 1998). However, a drawback of using PICT to detect adverse effects of soil contamination by TEs is the "natural" adaptation of the microbial community to geogenic sources of TEs (McLaughlin and Smolders 2001). At the same time, a lack of effect of long-term pollution by a TE on microbial communities may be due to adaptation of the community to the added TE, and function or population size may remain unaffected. Induced tolerance to TEs (whether through an anthropogenic source or a geogenic source) could be claimed to be an adverse effect if it impairs the organism (or function) in terms of imposing an energy cost, leading to lower resilience. There is evidence that, at least for Cu, Zn, and Pb, this does not occur (Rusk et al. 2004; Mertens et al. 2007b, 2010).

2.14.12 Mixture Considerations

It is acknowledged that ecosystems are generally exposed to mixtures of chemicals rather than to single toxicants; yet there is a paucity of mixture toxicity data, relative to that available for individual chemicals. The sheer number of possible mixtures, combined with the fact that changing the relative composition of a mixture (e.g., from 50% A and 50% B to 25% A and 75% B) can change its toxicity (e.g., European Inland Fisheries Advisory Committee 1980) makes any comprehensive assessment of mixture toxicity impossible. Therefore methods to predict the toxicity of mixtures have been developed. These extend back to the early classification schemes of Bliss (1939) and Placket and Hewlett (1952).

The Placket and Hewlett classification is presented in Table 2.4. Chemicals with the same mechanism of action (MeOA) (i.e., the same target site) and that do not interact have a simple similar mode of joint action that is also called concentration addition (CA) because the effect of the components of the mixture can substitute for each other. Each component of a simple similar mixture contributes to the toxicity, and thus they can be treated as a single toxicant. The combined effect of the components is equal to the sum of the concentrations of each chemical expressed as a fraction of its own individual toxicity. Chemicals with a dissimilar MeOA (i.e., different sites of action) that do not interact exert an independent mode of joint action, which is also called response addition (RA).

TABLE 2.4
Four types of joint action for mixtures developed by Plackett and Hewlett (1952)

	Similar joint action	Dissimilar joint action
Noninteractive	Simple similar (concentration addition, CA)	Independent (independent action, IA also called response addition, RA)
Interactive	Complex similar	Dependent

Source: Reprinted with permission from Dyer S, Warne MStJ, Meyer JS, Leslie HA, Escher BI. The tissue residue approach for chemical mixtures. IEAM DOI 10.1002/ieam.106.

For complex similar and dependent classes of joint action at least one chemical in the mixture affects the biological activity of at least one other chemical in the mixture. Chemicals in such mixtures may modify the biological activity of chemical "i" by affecting, for example, the rates of absorption, metabolism, or elimination of chemical "i," or by competing with chemical "i" for binding at the target site. When the chemicals in the mixture engaging in such interactions elicit toxicity through the same MeOA, the mixture toxicity is referred to as complex similar action. When the chemicals in the mixture act through different MeOAs, they are referred to as dependent joint action. In these latter 2 classes, synergism (the combined toxicity is greater than concentration or response addition) and antagonism (the combined toxicity is less than concentration or response addition) can be expected. Mathematical models have been developed for these interactions (Jonker et al. 2005; Dyer et al. forthcoming).

While the Plackett and Hewlett (1952) model has simplicity, it has a number of limitations (De Zwart and Posthuma, 2005). The main being that the MeOA of a chemical is not constant as it can vary with the duration of exposure, concentration, and species (Baird et al. 1990; Dyer et al. forthcoming). This variation in the MeOA that a chemical exerts means it is difficult to predict the MeOA of a chemical a priori.

The vast majority of the available literature on the toxicity of mixtures is for aquatic species rather than soil species, and thus they will form the basis of the following discussion. However, applying the results of aquatic mixture studies to soil species and to the derivation of SQSs requires a leap in faith and is based on the assumption that the principal exposure pathway is via soil solution.

Wang (1987) reviewed the acute toxicity of binary and complex TE mixtures and concluded that there is no consistent pattern nor could the toxicity be readily predicted. Despite this, 21 out of 37 mixtures examined were additive or antagonistic. A similar conclusion was reached for the chronic toxicity of mixtures of TEs by the European Centre for Ecotoxicology Toxicology of Chemicals (ECETOC) (2001). This unpredictability was thought to be due to the metals being examined having different mechanisms of action, test conditions, and test species (ECETOC 2001). In addition, some TEs are essential elements and may have more than one MeOA, and it is possible that each MeOA may have its own unique toxicity to organisms. For example, Braek et al. (1976) found that a binary mixture of Cu and Zn was synergistic or antagonistic, depending on the species of algae being tested. Researchers examining the toxicity of mixtures of TEs to soil organisms have also reported antagonism, additivity, slight synergism, and synergism (Korthals et al. 2000; Sneller et al. 2000; Odendaal and Reinecke 2004; Jonker et al. 2004; Shen et al. 2006). ECETOC (2001) therefore recommended that to assume "additivity is probably the most balanced choice, unless there is clear evidence in the literature that mixtures of the metals under examination behave differently" (p. 13).

Carlon (2007), in a review of soil screening values in 15 EU member nations, states that mixture effects are not considered in the screening risk assessment phase, i.e., derivation of SQS, with the exception of groups of compounds with the same MeOA (e.g., polycyclic aromatic hydrocarbons [PAHs]) which are defined by CA. The situation is similar in other jurisdictions such as the United States (USEPA 1996), Canada (CCME 2006), and Australia (National Environment Protection Council [NEPC] 1999).

The question arises then that given exposure is predominantly to mixtures, whether the current single chemical SQS provides adequate or the desired environmental protection? A series of laboratory-based studies (using organic chemicals) has shown that in the majority of mixtures containing chemicals with the same MeOA, the toxicity was consistent with CA, while mixtures of chemicals with different MeOAs exert toxicity consistent with RA (e.g., Hermens et al. 1984; Broderius and Kahl 1985; Hermens et al. 1985; Deneer et al. 1988; Broderius 1991; Altenburger et al. 1993; Faust et al. 1994; Broderius et al. 1995; Grimme et al. 1996; Van Leeuwen et al. 1996; Altenburger et al. 2000; Backhaus et al. 2000a, 2000b; Deneer 2000). Other studies (Faust et al. 1994; Warne and Hawker 1995; Ross and Warne 1997; Deneer 2000) that examined TE–TE mixtures, TE–organic chemical, and organic chemical–organic chemical mixtures found that approximately 5 to 15% of mixtures were antagonistic, 70 to 90% were additive, and 5 to 15% were synergistic. A more recent review by Norwood (2003) that focused exclusively on mixtures of TEs examined the aquatic toxicity of 191 mixtures to 77 species in 68 publications. Overall, they found that antagonistic, additive, and synergistic toxicity occurred in 43, 27, and 29% of the mixtures.

By combining the above information, the toxicity of the vast majority of mixtures would be estimated accurately or overestimated by CA, and only a small proportion of mixtures would have their toxicity underestimated. Further, aquatic toxicology papers by Faust et al. (1994), Backhaus et al. (2000a, 2000b), Dyer et al. (2000), Junghans et al. (2006), and Chevre et al. (2006) found that CA overestimated effects and yielded slightly higher estimates of the toxicity of mixtures than RA when chemicals had different MeOAs. The same result was found by Lock and Janssen (2002) for the terrestrial toxicity of TEs to the potworm *E. albidus*. Therefore the use of the CA model provides a conservative estimate of toxicity for mixtures that have a RA joint action.

Warne and Hawker (1995) developed the Funnel Hypothesis, which argues that the more components an equitoxic mixture (a mixture where each chemical is present at the same fraction of their individual toxicity e.g., TU of chemical A = measured concentration of chemical A/EC50 of chemical A) contains the more likely it is that the toxicity of the mixture will be consistent with CA (Figure 2.17).

Warne and Hawker (1995) collected aquatic toxicity data for 104 equitoxic mixtures composed of 182 chemicals including TEs and OCs to a diverse range of aquatic

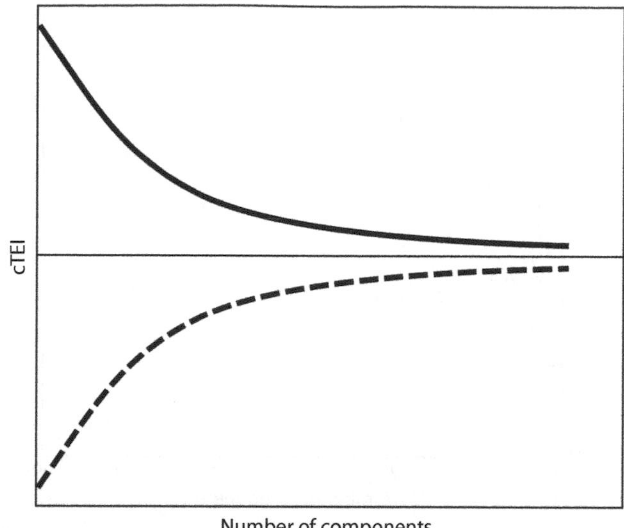

FIGURE 2.17 The variation in toxicity of mixtures with the number of components in the mixture predicted by the Funnel Hypothesis (reproduced from Ecotoxicology and Environmental Safety 31, Warne and Hawker, the number of components in a mixture determines whether synergistic, antagonistic or additive toxicity predominate: The Funnel Hypothesis, Copyright (2005), with permission Elsevier). The vertical axes is cTEI – the corrected Toxicity Enhancement Index which is a measure of the toxicity of mixtures. The horizontal line (cTEI = 0) corresponds to concentration addition. The two asymptotic lines are the maximum (top line) and minimum (bottom line) predicted toxicity of equitoxic mixtures as a function of the number of components. The more positive the cTEI value for a mixture the more synergistic the mixture. While the more negative the cTEI value for a mixture the more anatagonistic the mixture.

species. They found that these data conformed to the hypothesis (Figure 2.18). Ross (1996) and Ross and Warne (1997) found that toxicity data for 973 mixtures that included organic chemical–organic chemical, TE–TE, and TE–organic chemical mixtures, conformed to the predictions of the hypothesis. A limitation of the Funnel Hypothesis is that it only applies to equitoxic mixtures that are unlikely to occur in the environment. However, the underlying principle of the hypothesis means that its predictions should also apply to nonequitoxic mixtures. Thus 85 to 95% of chemical mixtures (including those involving TEs), for which there are data, are protected by assuming CA and as the number of chemicals in mixtures increases the toxicity of the mixture is increasingly likely to conform to CA. It is therefore important then to quantify the magnitude by which synergistic mixtures differ from CA. The analysis by Ross (1996) and Ross and Warne (1997) indicated that 5% of the mixtures had toxicity that differed from CA by a factor greater than 2.5, and only 1% of mixtures had toxicity values that differed from CA by more than a factor of 5.

The current practice of deriving chemical-specific SQSs can continue, provided mixtures are treated as though they conform to CA during the implementation phase. Whether the SQS are exceeded would then be calculated using the following formula:

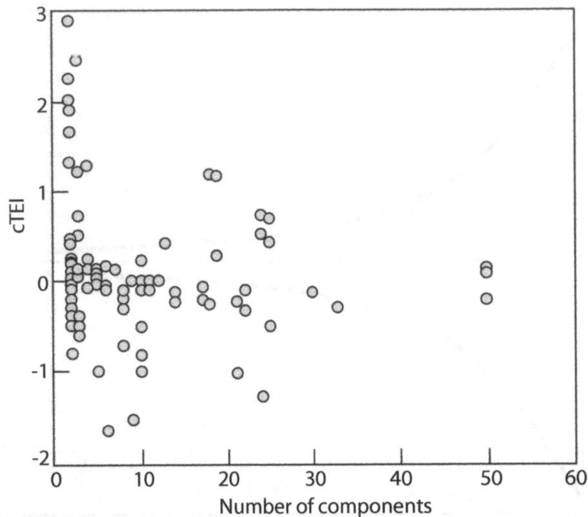

FIGURE 2.18 The observed variation in toxicity of mixtures with the number of components (reproduced from Ecotoxicology and Environmental Safety 31, Warne and Hawker, The number of components in a mixture determines whether synergistic, antagonistic or additive toxicity predominate: The Funnel Hypothesis, Copyright (2005), with permission Elsevier). The vertical axes is cTEI – the corrected Toxicity Enhancement Index which is a measure of the toxicity of mixtures. The horizontal line (cTEI = 0) corresponds to concentration addition. Corrected Toxicity Enhancement Index values of 2 and -2 mean that the actual toxicity of the mixtures is three times larger and one third as large, respectively, as if the mixture's toxicity conformed to concentration addition.

$$TTM = \Sigma(C_i/SQS_i),$$

where TTM is the total toxicity of the mixture, C_i is the concentration on the *i*th compound in the mixture, and SQS_i is the SQS for compound *i*. If TTM equals or is greater than 1, then the mixture has equaled or exceeded the SQS, respectively. If, however, the TTM is less than 1, then the SQS has not been exceeded.

De Zwart and Posthuma (2005) also showed how CA, RA, and the mixed model approaches could be combined with SSD methods so that the risk posed by mixtures could be assessed or potentially EQGs could be derived for mixtures.

It has been argued (e.g., Meador 2006) that much of the variation in toxicity values for a particular chemical across species (i.e., between the most and least sensitive species) is due to the fact that the toxicity data are expressed in terms of the ambient concentration (e.g., mg kg^{-1} soil) rather than the concentration at the target site. If this was the case, then it would also affect the classification of the toxicity of a mixture as antagonistic, additive, or synergistic. Thus some have argued that the toxicity of mixtures should be addressed using internal tissue concentrations (Borgmann et al. 2005; Dyer et al. forthcoming). Recent work by Kamo and Nagai (2008) modified the BLM so that it applied to mixtures. They found that when the toxicity of TEs was expressed in terms of aqueous concentrations, it could be antagonistic, additive, or synergistic, but when the toxicity was expressed in terms of bioavailable concentrations, the mixtures were all additive. They do state that their results are "still a theoretical outcome" and that it requires "toxicity testing using real organisms" (p. 1486). Borgmann et al. (2005) while not commenting on the work of Kamo and Nagai (2008) state that the BLM should "be used with caution when attempting to model metal interactions" (p. 266). With the development of terrestrial BLMs (e.g., Thakali et al. 2006a, 2006b; Lock et al. 2007a, 2007b; Luo et al. 2008), there is the potential for new methods of addressing the toxicity of mixtures into SQS to be developed.

Regardless of whether one believes that the findings of the aquatic mixture toxicity research apply to terrestrial ecosystems or that only the limited terrestrial mixture toxicity data are relevant makes no difference as the conclusion is the same. It is therefore recommended that mixture toxicity data continue to be treated using the CA model until more mechanistic models are developed that can accurately predict the toxicity of TE mixtures for a wide range of organisms.

2.14.13 Secondary Poisoning

Soil contamination with TEs directly affects organisms that live in the soil and depend on the soil ecosystem for nutrients (plants, soil invertebrates, and soil microbes). Wildlife (birds, mammals, reptiles, and adult amphibians) and aboveground invertebrates (Lepidoptera, Hymenoptera, and others) also can be affected by elevated TEs in soils through uptake in the food chain. A few countries have included wildlife protection in their SQSs, notably Belgium, Germany, Spain, Sweden, the United Kingdom (Carlon, 2007), Canada, and the United States, while other countries have not considered this aspect of the terrestrial ecosystem.

The derivation of SQSs for protection of wildlife is a 2-step process. First, information is needed on toxicity of TEs to wildlife, particularly effects on growth, reproduction, or survival (USEPA 2005). Data from chronic or subchronic feeding studies are preferred, with acute toxicity information (i.e., single oral or injected dose) considered less applicable to criteria development. Currently, the best sources of information on wildlife TE toxicity thresholds are National Academy of Sciences and National Research Council (1980, 1994), McDowell (2003), and the documentation supporting development of the USEPA Eco SSLs values (http://www.epa.gov/ecotox/ecossl/). Derivation of PNECs, or toxicity benchmarks and reference values, differs by country but most generally is based on either the lowest LOAEL or highest NOAEL available, although some jurisdictions (e.g., Canada) use effect doses such as the EC20. There is general agreement that bird and mammal PNECs should be derived separately, as avian and mammalian physiologies are sufficiently different so as to not group data from the 2 classes (Hull et al. 2007; Awkerman et al. 2008).

After the dietary exposure threshold (PNEC) values are established, corresponding soil concentrations are calculated using estimates of food ingestion rates (FIR), soil ingestion rates, and trophic transfer functions between soil and invertebrates, soil and plants, and plants and invertebrates (USEPA 2005; Carlon 2007). The exposure equation is further simplified by assuming that 100% bioavailability of TEs in food and soil and that the animals spend 100% of their life in the area of concern. In Europe, modifications to these assumptions have been made in more advanced tiers of assessment. For example, the bioavailability of TEs associated with food can be used if the initial worst case tier indicates risk of dietborne exposure. In the Ni secondary poisoning assessment for mammalian food chains, bioavailability was taken into account by applying a relative absorption factor, which was based on the ratio of absorption of Ni associated with food (which is relevant to what ecological receptors experience in nature) to the absorption of $NiSO_4$ (which is what the test organisms in the reference toxicity test were dosed) (Denmark 2008). Nearly always, the most protective soil concentration will be derived from modeling small animal exposures. Small animals such as shrews (*Sorex* or *Blarina* spp.) or songbirds have higher FIR per unit body weight than larger species and tend to have some of the highest incidental soil ingestion rates because of their burrowing and grooming habits and because they feed extensively on soil invertebrates (USEPA 2005). Trophic transfer functions for TEs for uptake from soil into invertebrates (Efroymson et al. 2001) or plants (Sample et al. 1999) indicate that, in general, soil invertebrates such as earthworms accumulate more TE than do plants, making the soil invertivore the most exposed wildlife species (Jongbloed 1996). In some jurisdictions, the assumption is made that diets of shrews and invertivore birds are 100% composed of earthworms. However, this can be a worst case assumption, and the assessment can be refined by using more accurate dietary compositions and by applying the appropriate trophic transfer function to each of the relevant dietary components. Estimating the proportion of various food items in wildlife diets can be a large source of uncertainty in SQS development. Shrews are assumed to eat primarily soil invertebrates (40 to 70% of their diet, depending on time of year), and small bodied birds such as the robin (*Turdus migratorius*) or bluebirds (*Sialia* spp.) also have a high proportion of soil invertebrates in their diets (McVey et al. 1993). Therefore these species are most frequently modeled to derive mammalian and avian SQS.

The SQSs developed for wildlife protection are highly uncertain owing to the variability of the input parameters. Of considerable importance is that TE concentrations in dietary items are a function of soil concentration (USEPA 2007a, 2007b), so a constant trophic transfer value may result in an inappropriately low soil concentration when SQSs are calculated from allowable dietary limits in food items. Assuming 100% bioavailability of TE in food items or from incidentally ingested soil is an overestimation of actual exposure, again adding conservatism to the final SQS. Consequently, the final SQS often is very low and may be below natural background levels in many areas (USEPA 2005).

There are currently insufficient data to develop SQSs for herpetofauna (amphibians and reptiles), which are assumed to be protected by the lowest value derived for birds and mammals. Some jurisdictions recommend use of a safety factor (generally 10) to provide additional protection to these animals in the absence of data (EC 2003). As there is no physiological or sound toxicological basis for making such extrapolations, and following strong recommendations by experts to never extrapolate toxicity data between classes of vertebrates (Hull et al. 2007), this practice should be discouraged.

Given all the uncertainties inherent in the derivation of SQSs for protection of wildlife, an alternative approach is to use tissue-specific critical loads for establishing critical exposure levels in wildlife. These have been calculated for some TEs, including Pb in liver, Cd in kidney, Hg in brains, and Se in eggs (Beyer et al. 1996).

The difficulties with this approach is the requirement to monitor wildlife in addition to soil and the fact that sampling many of the tissues requires invasive and frequently lethal procedures. Some TEs could be measured through sublethal means, such as blood levels (especially Pb and Zn), embryonated bird eggs (Se), or bird feathers (Hg and Pb). Regulatory actions would be triggered based on the results of the animal studies, not from soil concentration data.

2.15 CONCLUSIONS

This chapter has attempted to draw together the diverse information on TE chemistry in soils and the factors that need to be considered in determining the effective "dose," as well as the biotic factors that may affect the choice and use of ecotoxic "responses." Apart from political and social considerations, the variability in these factors helps to explain the widely diverse SQSs found across jurisdictions.

We have tried to present the state of the science in the preceding sections, with a view to providing guidance on the most appropriate information that is required to produce a TE SQS. There was much agreement among the authors at the workshop on the approaches and science that needs to be used to develop scientifically robust SQSs. Advancements have been made in this field in the past 5 years, largely as a result of the EU risk assessment process, and more robust and defensible SQSs are being developed as a result of the development of mechanistic or empirical models or as a result of better TE measurement methods.

2.15.1 MODELING

Mechanistic modeling has a conceptual underpinning that may help to justify its adoption with regulators, legislators, and the public. This is particularly pertinent in the case of the BLM, which is now being adopted for use in the aquatic environment in the United States (USEPA 2007a, 2007b), but only for an extremely limited number of TEs. Empirical and semiempirical modeling, while perhaps not as scientifically transparent, has the advantage of simplicity in derivation and application, and its value can be demonstrated by validation modeling (e.g., Denmark 2008). Both mechanistic and empirical modeling approaches also allow, in principle, for the continued use of historic toxicity data where the endpoint is expressed as a total or total added concentration of TE. This is a highly pertinent point for regulators as a number of environmentally important TEs, particularly the cationic metals such as Cd, Cu, Ni, Pb, and Zn have substantial historic toxicity data sets that allow guidelines to be developed using more sophisticated multispecies approaches, i.e., the use of SSDs. To use a direct measurement approach (e.g., various measures of pools of TEs in soil) would mean discarding almost all historic toxicity data and preclude, at least until sufficient data has been derived using these alternate measurements, the use of SSDs.

Although modeling approaches are potentially powerful tools, they possess a number of limitations. Ideally, a model derived from a particular data set should not be applied to soils whose composition lies outside this data set, without prior validation. This caveat applies equally to empirical and mechanistic models. Models are, by nature, specific to the species or process for which the effects of the TE of concern is being modeled, so models need to be applied carefully to data sets comprising other species and processes, and again validation is a key process. The validation process used in the Ni and Cu EU risk assessments are good examples of the testing of an empirical normalization model against available data. Further systematic validation of models across species, taxa, and phyla is a research priority. Modeling will also require additional soil composition parameters within both effects and exposure data sets. Ideally, such parameters would be routinely measured within exposure data sets or could be readily estimated from routine parameters, although it is possible that the choice of normalization model would have to be constrained by the available soil parameters and thus not be the most statistically optimal model possible.

Currently, the approach taken in the Cu, Ni, and Zn EU risk assessments (Smolders et al. 2009) uses empirical relationships similar to those listed in Table 2.2 to normalize effects data. It would be possible to use a parameterized tBLM in a similar manner. Careful consideration would need to be given to the data requirements of tBLMs, which might include competing cation concentrations (e.g., H, Na, Ca) and DOC in the soil solution. Measurements of the soil solution composition are not typically made for either effects or exposure data and not easily predicted from normal soil mapping or analysis programs, and so this would need to be considered for approaches that incorporate the tBLM. tBLMs could be developed that use solid phase measurements (e.g., exchangeable Na, Ca) in conjunction with measurements of soil binding phases (e.g., organic matter, clay, metal oxides) to predict soil solution free ion activities. Chemical speciation models that predict solid solution partitioning of TEs have been developed and tested against limited data sets (e.g., MacDonald and Hendershot 2006; Almas et al. 2007) with some success and offer promise for predicting soil solution chemistry from the soil solids, although their general applicability in a tBLM would need further validation.

Also central to the issue of the use of the tBLM is the development and validation of the biouptake (or biosorption) part of the model. While some researchers (e.g., Van Gestel and Koolhaas 2004; Antunes et al. 2007) have related effects to the uptake of TE by the target organism, many applications of the tBLM in the terrestrial environment have been made without reference to TE uptake, and the model is fitted in an essentially empirical manner without any understanding of the nature of metal uptake. On the one hand, this points to the flexibility of the tBLM, whereas, on the other hand, it is debatable whether it has any advantages over explicit empirical modeling, which might be able to describe similar effects data in a more statistically efficient manner. An improved understanding of the nature of TE uptake sites and internal pools in terrestrial organisms and their relationship with external TE and effects, would serve to improve confidence in the tBLM as a tool for use in soil guideline setting and risk assessment.

The modeling approach used in the Cu and Ni EU risk assessments appears to be a robust method of normalizing existing effects data. The same methodology has been adopted to derive SQSs in Flanders (VLAREBO 2008) and proposed for deriving SQSs in Australia (Heemsbergen et al. 2008; Warne et al. 2009). In contrast, the approach taken in the EU Zn risk assessment appears to be somewhat limited. Most importantly, in the EU Zn risk assessment there is no normalization of the effects (NOEC, ECx) data (effect side correction). As noted earlier, an effect value measured in a certain soil will be biased with respect to effects in a different soil and an SSD using nonnormalized effect values is not just a distribution of species sensitivity but also a distribution of soil bioavailability. It is therefore essential to normalize effect values to a common target soil composition prior to their use in the SSD if a robust enough way of doing this can be found. The normalization of exposure data only (exposure side correction) is not as robust a way of accounting for modifying factors because it does not account for effects of soil chemistry on effect values in toxicity tests.

2.15.2 Measurement

Direct measurement approaches (soil extractions) have the advantages of practicality and relative simplicity and have in a few cases already been incorporated into SQS setting frameworks (Carlon 2007). As was noted earlier, however, there is disagreement among researchers as to the best extraction for a particular TE, and it likely such disagreements arise from the fact that different extractants will extract different pools of TE from different soils, thus leading to inconsistent findings across different studies. Generally, weak salt solutions provide the best results (e.g., McLaughlin et al. 2000; Menzies et al. 2007), although the DGT technique shows promise for specific application to plants. Weak salt solutions probably extract the dissolved TE plus a small portion of solid phase complexed TE that desorbs in response to the chemical changes in the soil induced by the extraction (e.g., changes in pH, ionic strength, soil-solution ratio) (McLaughlin et al. 2000). Where knowledge is lacking is the extent to which the soil composition influences the amount of TE that can be extracted, and crucially, to what extent the extraction "mimics" the removal of TE from the soil or soil solution by organisms. In this respect the growing body of work on the use of DGT devices to estimate the "effective concentration" is interesting. DGT-collected TE appears to correlate well with TE uptake by plants, although there is little work yet done on its relationship with toxic effect or on its relationship with uptake by other organisms. Further testing and assessment of the approach is needed, including testing of its performance under field conditions.

It must be pointed out that there is an inherent contradiction between the tBLM and related models and the concept of extracting a TE from a soil using a weak salt solution as an indicator of its toxic effect. The tBLM explicitly states that the amount of TE at the site of toxic action is not a simple function of any single species of TE in the soil system but is rather a function of both TE chemical speciation and the competitive uptake of the TE and other ions at the biotic ligand. Empirical normalization models capture these chemical and biological processes in a single expression. More work is required on extraction methods to establish whether they can produce results directly relatable to toxic effect in a consistent manner across a range of soils. It would be particularly beneficial to compare modeling and measurement techniques within single studies to provide a direct comparison of their efficacy at describing effects. Such studies as have been carried out indicate that empirical modeling is a superior method, although further work is required to further compare the 2 approaches.

REFERENCES

Ahmed IAM, Young SD, Mosselmans JFW, Crout NMJ, Bailey EH. 2009. Coordination of Cd^{2+} ions in the internal pore system of zeolite-X: a combined EXAFS and isotopic exchange study. Geochim Cosmochim Acta 73(6):1577–1587.

Aldenberg T, Jaworska JS. 2000. Uncertainty of the hazardous concentration and fraction affected for normal species sensitivity distributions. Ecotoxicol Environ Saf 46:1–18.

Aldenberg T, Slob W. 1993. Confidence-limits for hazardous concentrations based on logistically distributed NOEC toxicity data. Ecotoxicol Environ Saf 25:48–63.

Almas AR, Lofts S, Mulder J, Tipping E. 2007. Solubility of major cations and Cu, Zn and Cd in soil extracts of some contaminated agricultural soils near a zinc smelter in Norway: modelling with a multisurface extension of WHAM. Eur J Soil Sci 58:1074–1086.

Almas AR, Lombnaes P, Sogn TA, Mulder J. 2006. Speciation of Cd and Zn in contaminated soils assessed by DGT-DIFFS, and WHAM/Model VI in relation to uptake by spinach and ryegrass. Chemosphere 62:1647–1655.

Altenburger R, Backhaus T, Boedeker W, Faust M, Scholze M, Grimme LH. 2000. Predictability of the toxicity of multiple chemical mixtures to *Vibrio fischeri*: mixtures composed of similarly acting chemicals. Environ Toxicol Chem 19:2341–2347.

Altenburger R, Boedeker W, Faust M, Grimme L. 1993. Aquatic toxicology, analysis of combination effects. In: Corn, M., editor, Handbook of hazardous materials. San Diego (CA): Academic Pr. p 15–27.

Alves BLR, Urquiaga S, Cadisch G, Souto CM, Boddy RM. 1993. In situ estimation of soil nitrogen mineralization. In: Mulongoy K, Merckx R., editors, Soil organic matter dynamics and sustainability of tropical agriculture. New York (NY): J Wiley. p 173–180.

Amacher MC, Selim HM, Iskandar IK. 1998. Kinetics of chromium(VI) and cadmium retention in soils; A nonlinear multireaction model. Soil Sci Soc Am J 52:398–408.

Amann R, Ludwig W, Schleifer KH. 1995. Phylogenetic identification and in situ detection of individual microbial cells without cultivation. Microbiol Rev 59:143–149.

Anderson JPE, Domsch KH. 1978. A physiological method for measurement of microbial biomass in soils. Soil Biol Biochem 10:215–221.

Antunes PMC, Berkelaar EJ, Boyle D, Hale BA, Hendershot W, Voigt A. 2006. The biotic ligand model for plants and metals: technical challenges for field application. Environ Toxicol Chem 25:875–882.

Antunes PMC, Hale BA. 2006. The effect of metal diffusion and supply limitations on conditional stability constants determined for durum wheat roots. Plant Soil 284:229–241.

Antunes PMC, Hale BA, Ryan AC. 2007. Toxicity versus accumulation for barley plants exposed to copper in the presence of metal buffers: progress towards development of a terrestrial biotic ligand model. Environ Toxicol Chem 26:2282–2289.

Australian and New Zealand Environment and Conservation Council and Agriculture and Resource Management Council of Australia and New Zealand (ANZECC and ARMCANZ). 2000. National water quality management strategy, Australian and New Zealand Guidelines for fresh and marine water quality. Canberra (Australia): ANZECC and ARMCANZ.

Awkerman JA, Raimondo S, Barron MG. 2008. Development of species sensitivity distributions for wildlife using interspecies toxicity correlation models. Environ Sci Technol 42(9):34–47.

Backhaus T, Altenburger R, Boedeker W, Faust M, Scholze M. 2000a. Predictability of the toxicity of a multiple mixture of dissimilarly acting chemicals to *Vibrio fischeri*. Environ Toxicol Chem 19:2348–2356.

Backhaus T, Altenburger R, Boedeker W, Faust M, Scholze M. 2000b. Predictability of the toxicity of a multiple mixture with combinations of environmental chemicals. Science 272:489–492.

Baird DJ, Barber I, Calow P. 1990. Clonal variation and general responses of *Daphnia magna* to toxic stress: 1. Chronic life history effects. Funct Ecol 4:399–407.

Barrow NJ. 1987. Reactions with variable charge soils. Dordrecht (Netherlands): Martinus Nijhoff Publ.

Barrow NJ, Gerth J, Brummer GW. 1989. Reaction kinetics of the adsorption and desorption of nickel, zinc and cadmium by goethite: II. Modelling the extent and rate of reaction. J Soil Sci 40:437–450.

BBodSchV, 1999. Bundes-Bodenschutz- und Altlastenverordnung (BBodSchV) vom 12. Juli 1999. (Federal Soil Protection and Contaminated Sites Ordinance dated 12 July 1999).

Beyer WN, Heinz GH, Redmon-Norwood AW. 1996. Environmnental contaminants in wildlife: interpreting tissue concentrations. Pensacola (FL): SETAC Pr., 494 p.

Bliss CI. 1939. The toxicity of poisons applied jointly. Ann Appl Biol 26:585–615.

Bongers M, Rusch B, Van Gestel CAM. 2004. The effect of counterion and percolation on the toxicity of lead for the springtail *Folsomia candida* in soil. Environ Toxicol Chem 23:195–199.

Bonnell Environmental Consulting. 1999. A proposed approach for estimating uncertainty factors for the ecological assessment of new substances in Canada. Final Report. Ontario (Canada): Commercial Chemicals Evaluation Branch. 61 p.

Borgmann U, Nowierski M, Dixon DG. 2005. Effect of major ions on the toxicity of copper to *Hyalella azteca* and implications for the biotic ligand model. Aquat Toxicol 73:268–287.

Braek GS, Jensen A, Mohus A. 1976. Heavy metal tolerance of marine phytoplankton: III Combined effects of copper and zinc on culture of four common species. J Exp Mar Biol Ecol 25:37–50.

Brix KV, DeForest DK, Adams WJ. 2001. Assessing acute and chronic copper risks to freshwater aquatic life using species sensitivity distributions for different taxonomic groups. Environ Toxicol Chem 20:1846–1856.

Broderius SJ. 1991. Modelling the joint toxicity of xenobiotics to aquatic organisms: basic concepts and approaches. In: Mayes MA, Barron MG, editors, Aquatic toxicology and risk assessment. 14th ed. Philadelphia (PA): American Society for Testing and Materials. p 107–127.

Broderius SJ, Kahl MD. 1985. Acute toxicity of organic chemical mixtures to the fathead minnow. Aquat Toxicol 6:307–322.

Broderius SJ, Kahl M, Hoglund MD. 1995. Use of joint toxic response to define the primary mode of toxic action for diverse industrial organic chemicals. Environ Toxicol Chem 14:1591–1605.

Broos K, Macdonald LM, Warne MSJ, Heemsbergen DA, Barnes MB, Bell M. McLaughlin MJ. 2007b. Limitations of soil microbial biomass carbon as an indicator of soil pollution in the field. Soil Biol Biochem 39(10):2693–2695.

Broos K, Mertens J, Smolders E. 2005. Toxicity of heavy metals in soil assessed with various soil microbial and plant growth assays: a comparative study. Environ Toxicol Chem 24:634–640.

Broos K, Warne MSJ, Heemsbergen DA, Stevens D, Barnes MB, Correll RL, McLaughlin MJ. 2007a. Soil factors controlling the toxicity of copper and zinc to microbial processes in Australian soils. Environ Toxicol Chem 26:583–590.

Bruemmer GW, Gerth J, Tiller KG. 1988. Reaction kinetics of the adsorption and desorption of nickel, zinc and cadmium by goethite: I. Adsorption and diffusion of metals. J Soil Sci 39:37–52.

Bundy JG, Paton GI, Campbell CD. 2004. Combined microbial community level and single species biosensor responses to monitor recovery of oil polluted soil. Soil Biol Biochem 36:1149–1159.

Burns G. 1977. Soil enzymology. Sci Prog 64:275–285.

Calabrese EJ, Baldwin LA 1993. Chemical-specific ecosystem MATC. In: Performing ecological risk assessments. Boca Raton (FL): Lewis Publ. p 165–183.

Campbell E, Palmer MJ, Shao Q, Warne M, Wilson D. 2000. BurrliOZ: A computer program for calculating toxicant trigger values for the ANZECC and ARMCANZ water quality guidelines. The Australian and New Zealand Environment and Conservation Council (ANZECC) and Australian Resource Management Council of Australia and New Zealand (ARMCANZ), Canberra, ACT, Australia.

Campbell PGC. 1995. Interactions between trace metals and aquatic organisms: a critique of the free ion activity model. In: Tessier A, Turner DR, editors. Metal speciation and bioavailability in aquatic systems. Vol. 2. New York: J Wiley. p 45–97.

Carlon C, editor. 2007. Derivation methods of soil screening values in Europe. A review and evaluation of national procedures towards harmonization. Report EUR 22805-EN Ispra (Italy): European Commission, Joint Research Centre. 306 p.

Canadian Council for Ministers for the Environment (CCME). 1996. A protocol for the derivation of environmental and human health soil quality guidelines. ISBN-10 1-896997-45-7 PDF. ISBN-13 978-1-896997-45-2 PDF. Winnipeg, Manitoba.

Canadian Council for Ministers for the Environment (CCME). 2006. A protocol for the derivation of environmental and human health soil quality guidelines. Winnipeg, Manitoba Canada, 215 p. http://www.ccme.ca/assets/pdf/sg_protocol_1332_e.pdf (accessed 21 March 2009).

Chapman PM, Cardwell RS, Chapman PF. 1996. A warning: NOECs are inappropriate for regulatory use. Environ Toxicol Chem 15(2):77–79.

Chapman PM, McDonald BG, Kickham PE, McKinnon S. 2006. Global geographic differences in marine metals toxicity. Mar Pollut Bull 52(9):1081–1084.

Chaudri AM, McGrath SP, Giller KE. 1992. Survival of the indigenous population of *Rhizobium leguminosarum* biovar *trifolii* in soil spiked with Cd, Zn, Cu and Ni salts. Soil Biol Biochem 24:625–632.

Chèvre N, Loepfe C, Fenner K, Singer H, Escher BI, Stamm C. 2006. Wirkungsbasierte Qualitätskriterien für Pestizide in Oberflächengewässern der Schweiz. Ein konsistentes Konzept für Einzelstoffe und Mischungen. Gas Wasser Abwasser 4: 297–307.

Conder JM, Lanno RP. 2000. Evaluation of surrogate measures of cadmium, lead, and zinc bioavailability to *Eisenia fetida*. Chemosphere 41:1659–1668.

Cornu JY, Denaix L. 2006. Prediction of zinc and cadmium phytoavailability within a contaminated agricultural site using DGT. Environ Chem 3:61–64.

Criel P, Lock K, Van Eeckhout H, Oorts K, Smolders E, Janssen CR. 2008. Influence of soil properties on copper toxicity for two soil invertebrates. Environ Toxicol Chem 27:1748–1755.

Crommentuijn T, Polder M, Sijm D, de Bruijn J, van de Plassche E. 2000a. Evaluation of the Dutch environmental risk limits for metals by application of the added risk approach. Environ Toxicol Chem 19(6):1692–1701.

Crommentuijn T, Sijm D, de Bruijn J, van den Hoop M, van Leeuwen K, van de Plassche E. 2000b. Maximum permissible and negligible concentrations for metals and metalloids in the Netherlands, taking into account background concentrations. J Environ Manag 60:121–143.

Crout NMJ, Tye AM, Zhang H, McGrath SP, Young SD. 2006. Kinetics of metal fixation in soils: measurement and modeling by isotopic dilution. Environ Toxicol Chem 25(3):659–663.

De Schamphelaere KAC, Janssen CR. 2002. A biotic ligand model predicting acute copper toxicity for *Daphnia magna*: the effects of calcium, magnesium, sodium, potassium, and pH. Environ Sci Technol 36:48–54.

De Zwart D, Posthuma L. 2005. Complex mixture toxicity for single and multiple species: proposed methodologies. Environ Toxicol Chem 24(10):2665–2676.

Degryse F, Smolders E, Merckx R. 2006a. Labile Cd complexes increase Cd availability to plants. Environ Sci Technol 40:830–836.

Degryse F, Smolders E, Parker DR. 2006b. An agar gel technique demonstrates diffusion limitations to cadmium uptake by higher plants. Environ Chem 3:419–423.

Degryse F, Smolders E, Zhang H, Davison W. 2009. Predicting availability of mineral elements to plants with the DGT technique: a review of experimental data and interpretation by modelling. Environ Chem 6:198–218.

Deneer JW. 2000. Toxicity of mixtures of pesticides in aquatic systems. Pest Manag Sci 56:516–520.

Deneer JW, Sinnige TL, Seinen W, Hermens JLM. 1988. The joint acute toxicity to *Daphnia magna* of industrial organic chemicals at low concentrations. Aquat Toxicol 12:33–38.

Denmark. 2008. European Union risk assessment report on nickel, nickel sulphate, nickel carbonate, nickel chloride, nickel dinitrate. Report prepared by Denmark, Danish Environmental Protection Agency on behalf of the European Union.

Dijkstra JJ, Meeussen JCL, Comans RNJ. 2004. Leaching of heavy metals from contaminated soils: an experimental and modeling study. Environ Sci Technol 38:4390–4395.

Dilly O, Kutsch WL. 2000. Rationale for aggregating microbiological information for soil and ecosystem research. In: Benedetti A, Tittarelli F, Bertoldi S, Pinazari F., editors. Biotechnology of soil: monitoring, conservation and remediation, EUR 19548. Brussels: European Commission. p 135–140.

Donner E, Broos K, Heemsbergen D, Warne M, McLaughlin MJ, Hodgson ME, Nortcliff S. 2010. Biological and chemical assessment of zinc ageing in field soils. Environ Pollut 158:339–345.

Dyer S, Warne M, Meyer J, Leslie HA, Escher B. Forthcoming. Tissue residues of chemical mixtures. IEAM.

Dyer SD, Belanger SE, Carr GJ. 1997. An initial evaluation of the use of Euro/North American fish species for tropical effects assessments. Chemosphere 35(11):2767–2781.

Dyer SD, White-Hull CE, Shepard BK. 2000. Assessments of chemical mixtures via toxicity reference values overpredict hazard to Ohio fish communities. Environ Sci Technol 34:2518–2524.

Dzombak DA, Morel FMM. 1990. Surface complexation modeling: hydrous ferric oxide. New York: Wiley-Interscience. 393 p.

Efroymson RA, Sample BE, Suter GW. 2001. Bioaccumulation of inorganic chemicals from soil by plants. Report prepared for the U.S. Department of Energy under contract DE-AC05-84OR21400. Code EW 20. Oak Ridge (TN): Oak Ridge National Laboratory.

Emans HJB, Van de Plassche EJ, Canton JH, Okkerman PC, Sparenburgs PM. 1993. Validation of some extrapolation methods for effects assessment. Environ Toxicol Chem 12:2139–2154.

Euras. 2006. Terrestrial effects assessment of Ni. Report prepared by P. Van Sprang, Euras, Ghent, Belgium for the Nickel Producers Environmental Research Association, Durham (NC). 120 p.

European Commission (EC). 2003. Technical guidance document on risk assessment in support of Commission Directive 93/67/EEC on Risk Assessment for new notified substances, Commission Regulation (EC) No 1488/94 on Risk Assessment of existing substances and Directive 98/8/EC of the European Parliament and of the Council concerning the placing of biocidal products on the market. Report EUR 20418 EN. Ispra, Italy: Joint Research Centre.

European Chemicals Agency (ECHA). 2008. Guidance on information requirements and chemical safety assessment. Appendix R.7.13-2. Environmental risk assessment for metals and metal compounds. Helsinki (Finland): ECHA.

European Chemicals Agency (ECHA). 2009. Guidance on information requirements and chemical safety assessment. Chapter R.16: Environmental exposure estimation. Draft Version 2.0 (23.10.09). Helsinki (Finland): ECHA.

European Centre for Ecotoxicology and Toxicology of Chemicals (ECETOC). 2001. Aquatic toxicity of mixtures. Technical Report 80. Brussels (Belgium): ECETOC. 64 p.

European Copper Institute. 2008. Voluntary Risk Assessment of copper, copper(II) sulphate pentahydrate, copper(I) oxide, copper(II) oxide, dicopper chloride trihydroxide. Brussels (Belgium): European Copper Institute.

European Inland Fisheries Advisory Committee. 1980. Combined effects on freshwater fish and other aquatic life of mixtures of toxicants in water. Rome (Italy): Food and Agriculture Organisation of the United Nations.

Faust M, Altenberger R, Boedeker W, Grimme LH. 1994. Algal toxicity of binary combinations of pesticides. Bull Environ Contam Toxicol 53:134–141.

Fischer L, Zur Muhlen E, Brümmer GW, Niehus H. 1996. Atomic force microscopy (AFM) investigations of the surface topography of a multidomain porous goethite. Eur J Soil Sci 47:329–334.

Forbes TL, Forbes VE. 1993. A critique of the use of distribution-based extrapolation models in ecotoxicology. Funct Ecol 7:249–254.

Forbes VE, Calow P. 2002. Species sensitivity distributions revisited. A critical appraisal. Hum Ecol Risk Assess 8(3):473–492.

Fox DR. 2008. NECs, NOECs and the ECx. Australas J. Ecotoxicol 14:7–9.

Fountain MT, Hopkin SP. 2004. A comparative study of the effects of metal contamination on Collembola in the field and in the laboratory. Ecotoxicology 13:573–587.

Fritze H, Vanhala P, Pietikäinen J, Mälkönen E. 1996. Vitality fertilization of Scots pine stands growing along a gradient of heavy metal pollution: short-term effects on microbial biomass and respiration rate of the humus layer. Fresenius J Anal Chem 354:750–755.

Giller KE, Witter E, McGrath SP. 1999. Assessing risks of heavy metal toxicity in agricultural soils: do microbes matter? Hum Ecol Risk Assess 5(4):683–689.

Giller KE, Witter E, McGrath SP. 2009. Heavy metals and soil microbes. Soil Biol Biochem 41:2031–2037.

Gimmler H, de Jesus J, Greiser A. 2001. Heavy metal resistance of the extreme acidotolerant filamentous fungus *Bispora* sp. Microb Ecol 42:87–98.

Grimme LH, Faust M, Boedeker W, Altenburger R. 1996. Aquatic toxicity of chemical substances in combination: still a matter of controversy. Hum Ecol Risk Assess 2:426–433.

Grist EPM, Leung KMY, Wheeler JR, Crane M. 2002. Better bootstrap estimation of hazardous concentration thresholds for aquatic assemblages. Environ Toxicol Chem 21:1515–1524.

Gupta SK, Aten C. 1993. Comparison and evaluation of extraction media and their suitability in a simple model to predict the biological relevance of heavy metal concentrations in contaminated soils. Int J Environ Anal Chem 51:25–46.

Gupta SK, Vollmer MK, Krebs R. 1996. The importance of mobile, mobilisable and pseudo total heavy metal fractions in soils for three-level risk assessment and risk management. Sci Tot Environ 178:11–20.

Gustafsson JP. 2006. Arsenate adsorption to soils: modelling the competition from humic substances. Geoderma 136:320–330.

Hamon RE, McLaughlin MJ, Gilkes RJ, Rate AW, Zarcinas B, Robertson A, Cozens G, Radford N, Bettenay L. 2004. Geochemical indices allow estimation of heavy metal background concentrations in soils. Global Biogeochem Cycles 18(GB1014):1–6.

Hatch DJ, Jarvis SC, Reynolds E. 1991. An assessment of the contribution of net mineralization to N cycling in grass swards using a field incubation method. Plant Soil 138:23–32.

He Z, Gentry TJ, Schadt CW, Wu L, Liebich J, Chong SC, Huang Z, Wu W, Gu B, Jardine P, Criddle C, Zhou J. 2007. GeoChip: a comprehensive microarray for investigating biogeochemical, ecological and environmental processes. ISME J 1:67–77.

Heemsbergen D, Warne M, McLaughlin M, Kookana R. 2008. Proposal for an Australian methodology to derive ecological investigation levels in contaminated soils. Science Report 15/08. Adelaide (Australia): CSIRO Land and Water. 77 p.

Heemsbergen DA, Warne M, Broos K, Bell M, Nash D, McLaughlin M, Whatmuff M, Barry G, Pritchard D, Penney N. 2009a. Application of phytotoxicity data to a new Australian soil quality guideline framework for biosolids. Sci Total Environ 407(8): 2546–2556.

Heemsbergen DA, Warne M, McLaughlin MJ, Kookana R. 2009b. Proposed ecological investigation levels for arsenic, DDT, naphthalene and zinc. Science Report 02/09. Adelaide (Australia): CSIRO Land and Water. 89 p.

Hermens JLM, Canton H, Janssen P, de Jong R. 1984. QSARs and toxicity studies of mixtures of chemicals with anaesthetic potency: acute lethal and sub-lethal toxicity to *Daphnia magna*. Aquat Toxicol 5:143–154.

Hermens JLM, Leeuwangh P, Musch A. 1985. Joint toxicity of mixtures of groups of organic aquatic pollutants to the guppy (*Poecilia retiuclata*). Ecotoxicol Environ Saf 9:321–326.

Hickey GL, Kefford BJ, Dunlop JE, Craig PS. 2008. Making species sensitivity distributions reflective of naturally occurring communities: using rapid testing and Bayesian statistics. Environ Toxicol Chem 27(1):2403–2411.

Hobbs DA, Warne M, Markich SJ. 2004. Utility of Northern Hemisphere metal toxicity data in Australasia. SETAC Globe 5(2):38–39.

Hoekstra JA, Van Ewijk PH. 1993. Alternatives for the no-observed effect level. Environ Toxicol Chem 12:187–194.

Hose GC, Van den Brink PJ. 2004. Confirming the species-sensitivity distribution concept for endosulfan using laboratory, mesocosm, and field data. Arch Environ Contam Toxicol 47(4):511–520.

Hull RN, Allard P, Fairbrother A, Hope B, Johnson MS, Kapustka LA, McDonald B, Sample BE. 2007. Development and use of wildlife TRVs in ecological risk assessment. Extended abstracts. SETAC North America Conference, Nov. 2007, Milwaukee (WI). Available from http://milwaukee.setac.org/milwaukee/general/downloadable.php. Accessed 15 February 2009.

International Organization for Standardization. 2005. Soil quality: guidance on the determination of background values. Geneva (Switzerland): ISO. (ISO CD 19258).

Jansch S, Amorim MJ, Rombke J. 2005. Identification of the ecological requirements of important terrestrial ecotoxicological test species. Environ Rev 13:51–83.

Jenkinson DS, Hart PBS, Rayner JN, Parry LC. 1987. Modelling the turnover of organic matter in long-term experiments at Rothamsted. INTECOL Bull. 15:1–8.

Jenkinson DS, Ladd JN. 1981. Microbial biomass in soil: measurement and turnover. In: Paul EA, Ladd JN, editors. Soil biochemistry. Vol. 5. New York: Dekker. p 415–471.

Jenkinson DS, Powlson DS. 1976. The effect of biocidal treatment on metabolism in soil. V. A method for measuring soil biomass. Soil Biol Biochem 8:209–213.

Jongbloed RH, Traas TP, Luttik R. 1996. A probabilistic model for deriving soil quality criteria based on secondary poisoning of top predators: II. Calculations for dichlorodiphenyltrichloroethane (DDT) and cadmium. Ecotoxicol Environ Saf 34:279–306.

Jonker MJ, Svendsen C, Bedaux JJM, Bongers M, Kammenga JE. 2005. Significance testing of synergistic/antagonistic, dose level-dependent, or dose ratio-dependent effects in mixture dose-response analysis. Environ Toxicol Chem 24(10):2701–2713.

Jonker MJ, Sweijen RAJC, Kammenga JE. 2004. Toxicity of simple mixtures to the nematode *Caenorhabditis elegans* in relation to soil sorption. Environ Toxicol Chem 23(2):480–488.

Junghans M, Backhaus T, Faust M, Scholze M, Grimme LH. 2006. Application and validation of approaches for the predictive hazard assessment of realistic pesticide mixtures. Aquat Toxicol 76:93–110.

Kalisz HM. 1988. Microbial proteases. Adv Biochem Eng Biotechnol 36:1–65.

Kamo M, Nagai T. 2008. An application of the biotic ligand model to predict the toxic effects of metal mixtures. Environ Toxicol Chem 27:1479–1487.

Kapustka LA. 1999. Microbial endpoints: the rationale for their exclusion as ecological assessment endpoints. Hum Ecol Risk Assess 5(4):691–696.

Kefford BJ, Nugegoda D, Metzeling L, Fields EJ. 2006. Validating species sensitivity distributions using salinity tolerance of riverine macroinvertrebrates in the southern Murray–Darling Basin (Victoria, Australia). Can J Fish Aquat Sci 63:1865–1877.

Kinraide TB. 2003. The controlling influence of cell-surface electrical potential on the uptake and toxicity of selenate (SeO_4^{2-}). Physiol Plant 117:64–71.

Kinraide TB, Pedler JF, Parker DR. 2004. Relative effectiveness of calcium and magnesium in the alleviation of rhizotoxicity in wheat induced by copper, zinc, aluminum, sodium, and low pH. Plant Soil 259:201–208.

Kirk JL, Beaudette LA, Hart M, Moutoglis P, Khironomos JN, Lee H, Trevors JT. 2004. Methods of studying soil microbial diversity. J Microbiol Methods 58:169–188.

Kooijman SALM. 1987. A safety factor for LC50 values allowing for differences in sensitivity among species. Water Res 21:269–276.

Korthals GW, Bongers M, Fokkema A, Dueck TA, Lexmond TM. 2000. Joint toxicity of copper and zinc to a terrestrial nematode community in an acid sandy soil. Ecotoxicology 9(3):219–228.

Koster M, de Groot A, Vijver M, Peijnenburg WJGM. 2006. Copper in the terrestrial environment: verification of a laboratory-derived terrestrial biotic ligand model to predict earthworm mortality with toxicity observed in field soils. Soil Biol Biochem 38:1788–1976.

Koster M, Reijnders L, van Oost NR, Peijnenburg W. 2005. Comparison of the method of diffusive gels in thin films with conventional extraction techniques for evaluating zinc accumulation in plants and isopods. Environ Pollut 133:103–116.

Kwok KWH, Leung KMY, Chu VKH, Lam PKS, Morritt D, Maltby L, Brock TCM, Van den Brink PJ, Warne M, Crane M. 2007. Comparison of tropical and temperate freshwater species sensitivities to chemicals: implications for deriving safe extrapolation factors. Integrated Environ Assess Manag 3(1):49–67.

Ladd JN, Butler JHA. 1972. Short-term assays of soil proteolytic enzyme activities using proteins and dipeptide derivates as substrates. Soil Biol Biochem 4:19–30.

Lebourg A, Sterckman T, Ciesielski H, Proix N. 1996. Intérêt de différents réactifs d'extraction chimique pour l'évaluation de la biodisponibilité des métaux en traces desol. Agronomie 16:201–215.

Lexmond TM, Edelman T. 1994. Huidige achtergrondwaarden van het gehalte ann een aantal zware metalen en arseen in grond. In: de Haan FAM, Henkens CH, Zeilmaker DA, editors. Bodembescherming. Handboek Voor Milieubeheer. Wageningen (Netherlands): Samson H.D. Tjeenk Willink bv, Alphen aan den Rijn. p D4110-1-D4110-35.

Li HF, Gray C, Mico C, Zhao FJ, McGrath SP 2009. Phytotoxicity and bioavailability of cobalt to plants in a range of soils. Chemosphere 75:979–986.

Lock K, Criel P, De Schamphelaere KAC, Van Eeckhout H, Janssen CR. 2007b. Influence of calcium, magnesium, sodium, potassium and pH on copper toxicity to barley (*Hordeum vulgare*). Ecotoxicol Environ Saf 68(2):299–304.

Lock K, De Schamphelaere KAC, Becaus S, Criel P, Van Eeckhout H, Janssen CR. 2006. Development and validation of an acute biotic ligand model (BLM) predicting cobalt toxicity in soil to the potworm *Enchytraeus albidus*. Soil Biol Biochem 38: 1924-1932.

Lock K, De Schamphelaere KAC, Becaus S, Criel P, Van Eeckhout H, Janssen CR. 2007a. Development and validation of a terrestrial biotic ligand model predicting the effect of cobalt on root growth of barley (*Hordeum vulgare*). Environ Pollut 147:626–633.

Lock K, Janssen CR. 2001. Ecotoxicity of Zn in spiked artificial soils versus contaminated field soils. Environ Sci Technol. 35:4295–4300.

Lock K, Janssen CR. 2002. Mixture toxicity of zinc, cadmium, copper, and lead to the potworm *Enchytraeus albidus*. Ecotoxicol Environ Saf 52(1):1–7.

Lock K, Janssen CR, de Coen WM. 2000. Multivariate test designs to assess the influence of zinc and cadmium bioavailability in soils on the toxicity to *Enchytraeus albidus*. Environ Toxicol Chem 19:2666–1671.

Lofts S, Spurgeon D, Svendsen C, Tipping E. 2004. Deriving soil critical limits for Cu, Zn, Cd, and Pb: a method based on free ion concentrations. Environ Sci Technol 38:3623–3631.

Luo XS, Li LZ, Zhou DM. 2008. Effect of cations on copper toxicity to wheat root: implications for the biotic ligand model. Chemosphere 73(3):401–406.

Ma YB, Lombi E, Oliver IW, Nolan AL, McLaughlin MJ. 2006. Long-term aging of copper added to soils. Environ Sci Technol 40: 6310–6317.

Maas EV. 1986. Salt tolerance of plants. Appl Agric Res 1(1):12–26.

MacDonald JD, Belanger N, Hendershot WH. 2004a. Column Leaching using dry soil to estimate solid-solution partitioning observed in zero-tension lysimeters: 1. Method development. Soil Sed Contam 13:361–374.

MacDonald JD, Belanger N, Hendershot WH. 2004b. Column Leaching using dry soil to estimate soilid-solution partitioning observed in zero-tension lysimeters. 2. Trace metals. Soil Sed Contam 13:375–390.

MacDonald JD, Hendershot WH. 2006. Modelling trace metal partitioning in forest floors of northern soils near metal smelters. Environ Pollut 143:228–240.

Maltby L, Blake N, Brock TCM, Van den Brink PJ. 2002. Addressing interspecific variation in sensitivity and the potential to reduce this source of uncertainty in ecotoxicological assessments. Science and Research Report PN0932. London (UK): Department for Environment, Food and Rural Affairs. Also available from http://randd.defra.gov.uk/Document.aspx?Document=PN0932_1148_FRP.pdf.

Maltby L, Blake N, Brock TCM, Van den Brink PJ. 2005. Insecticide species sensitivity distributions: importance of test species selection and relevance to aquatic ecosystems. Environ Toxicol Chem 24(2):379–388.

Markich SJ, Camilleri C. 1997. Investigation of metal toxicity to tropical biota: Recommendations for revision of the Australian water quality guidelines. Report 127. Canberra (Australia): Supervising Scientist.

Martens R. 1995. Current methods for measuring microbial biomass C in soil: potentials and limitations. Biol Fertil Soils 19:87–99.

McDowell LR. 2003. Minerals in animal and human nutrition. 2nd ed. Amsterdam (Netherlands): Elsevier.

McLaughlin MJ. 2001a. Ageing of metals in soils changes bioavailability. London (UK): International Council on Mining and Metals, http://www.icmm.com/page/1345/envi-romental-fact-sheet-5-ageing-of-metals-in-soils-changes-bioavailability (accessed 18 September 2009).

McLaughlin MJ. 2001b. Bioavailability of metals to terrestrial plants. In: Allen H, editor. Bioavailability of metals in terrestrial ecosystems: importance of partitioning for bioavailability to invertebrates, microbes and plants. Pensacola (FL): Society of Environmental Toxicology and Chemistry (SETAC). p 39–68.

McLaughlin MJ. 2002. Heavy metals. In Lal R, editor. Encyclopedia of soil science. New York (NY): Marcel Dekker.

McLaughlin M, Hamon R, Parker DR, Pierzynski GM, Smolders E, Thornton I, Welp G. 2002. Test methods to determine hazards of sparingly soluble metal compounds in soils. Pensacola (FL): SETAC.

McLaughlin MJ, Smolders E. 2001. Background zinc concentrations in soil affect the zinc sensitivity of soil microbial processes—a rationale for a metalloregion approach to risk assessments. Environ Toxicol Chem 20(11):2639–2643.

McLaughlin MJ, Smolders E, Merckx R, Maes A. 1997. Plant uptake of cadmium and zinc in chelator-buffered nutrient solution depends on ligand type. In: Ando T, Fujita K, Mae T, Matsumoto H, Mori S, Sekija J, editors. Plant nutrition for sustainable food production and environment. Dordrecht (Netherlands): Kluwer Academic. p 113–118.

McLaughlin MJ, Whatmuff M, Warne M, Heemsbergen D, Barry G, Bell M, Nash D, Pritchard D. 2006. A field investigation of solubility and food chain accumulation of biosolid-cadmium across diverse soil types. Environ Chem 3:428–432.

McLaughlin MJ, Zarcinas BA, Stevens DP, Cook N. 2000. Soil testing for heavy metals. Commun Soil Sci Plant Anal 31(11–14):1661–1700.

McVey M, Hall K, Trenham P, Soast A, Frymier L, Hirst A. 1993. Wildlife exposure factors handbook. Vol. I. Report EPA/600/R-93/187a. Environmental Protection Agency, Washington (DC): USEPA.

Meador J. 2006. Rationale and procedures for using the tissue-residue approach for toxicity assessment and determination of tissue, water and sediment quality guidelines for aquatic organisms. Hum Ecol Risk Assess 12:1018–1073.

Menzies NW, Donn MJ, Kopittke PM. 2007. Evaluation of extractants for estimation of the phytoavailable trace metals in soils. Environ Pollut 145:121–130.

Mertens J, Springael D, De Troyer I, Cheyns K, Wattiau P, Smolders E. 2006. Long-term exposure to elevated zinc concentrations induced structural changes and zinc tolerance of the nitrifying community in soil. Environ Microbiol 8(12):2170–2178.

Mertens J, Degryse F, Springael D, Smolders E. 2007a. Zinc toxicity to nitrification in soil and soilless culture can be predicted with the same biotic ligand model. Environ Sci Technol 41:2992–2997.

Mertens J, Ruyters S, Springael D, Smolders E. 2007b. Resistance and resilience of zinc tolerant nitrifying communities is unaffected in long-term zinc contaminated soils. Soil Biol Biochem 39(7):1828–1831.

Mertens J, Wakelin SA, Broos K, McLaughlin MJ, Smolders E. 2010. Extent of copper tolerance and consequences for functional stability of the ammonia-oxidizing community in long-term copper-contaminated soils. Environ Toxicol Chem 29: 27–37.

Mico C, Li HF, Zhao FJ, McGrath SP. 2008. Use of Co speciation and soil properties to explain variation in Co toxicity to root growth of barley (*Hordeum vulgare* L.) in different soils. Environ Pollut 156:883–890.

Ministry of Environmental Protection of the People's Republic of China. 1995. Environmental quality standard for soils. Report GB15618-1995. Beijing (China). Also available from: http://english.mep.gov.cn/standards_reports/standards/Soil/Quality_Standard3/200710/W020070313485587994018.pdf

Ministry of Housing, Spatial Planning and Environment. 2000. Circular on target values and intervention values for soil remediation. Ref. DBO/1999226863. Bilthoven (Netherlands): Ministry of Housing, Spatial Planning and the Environment. 61 p.

Mortvedt JJ, Giordano PM. 1969. Extractability of zinc granulated with macronutrient fertilizers in relation to its agronomic effectiveness. J Agric Food Chem 17(6):1272–1275.

Muller AK, Westergaard K, Christensen S, Sorensen SJ. 2001. The effect of long-term mercury pollution on the soil microbial community. FEMS Microbiol Ecol 36:11–19.

National Academy of Sciences/National Research Council (US). 1980. Mineral tolerance of domestic animals. Washington (DC): National Academic Pr.

National Academy of Sciences/National Research Council (US). 1994. Nutritional requirements of poultry. 9th ed. Washington (DC): National Academic Pr.

National Environment Protection Council (NEPC). 1999. National Environment Protection (Assessment of Site Contamination) Measure 1999. Schedule B(1) Guideline on the investigation levels for soil and groundwater. Adelaide (Australia): NEPC. 16 p.

Netherlands. 2004. 2004 European Union risk assessment report on zinc metal, zinc(ii)chloride, zinc sulphate, zinc distearate, zinc oxide, trizinc bis(orthophosphate). Report prepared for RIVM on behalf of the European Union.

Newman MC. 2008. What exactly are you inferring? A closer look at hypothesis testing. Environ Toxicol Chem 27:1013–1019.

Newman MC, Ownby DR, Mezin LCA, Powell DC, Christensen TRL, Lerberg SB, Anderson BA. 2000. Applying species sensitivity distributions in ecological risk assessment: assumptions of distribution type and sufficient number of species. Environ Toxicol Chem 19:508–515.

Nolan AL, Zhang H, McLaughlin MJ. 2005. Prediction of zinc, cadmium, lead, and copper availability to wheat in contaminated soils using chemical speciation, diffusive gradients in thin films, extraction, and isotopic dilution techniques. J Environ Qual 34:496–507.

Noppert F, Van der Hoeven N, Leopold A. 1994. How to measure no effect? Towards a new measure of chronic toxicity in ecotoxicology. Delft: Netherlands Working Group on Statistics and Ecotoxicology.

Norwood WP, Borgmann U, Dixon DG, Wallace A. 2003. Effects of metal mixtures on aquatic biota: a review of observations and methods. Hum Ecol Risk Assess 9(4):795–811.

Nowack B, Koehler S, Schulin R. 2004. Use of diffusive gradients in thin films (DGT) in undisturbed field soils. Environ Sci Technol 38:1133–1138.

Odendaal JP, Reinecke AJ. 2004. Effect of metal mixtures (Cd and Zn) on body weight in terrestrial isopods. Arch Environ Contam Toxicol 46(3):377–384.

Organisation for Economic Co-operation and Development (OECD). 1992. Report of the OECD Workshop on extrapolation of laboratory aquatic toxicity data to the real environment. Environment Monographs 59. Paris (France): OECD.

Organisation for Economic Co-operation and Development (OECD). 1995. Guidance document for aquatic effects assessment. Environment Monographs 92. Paris (France): OECD.

Okkerman PC, Van de Plassche EJ, Emans HJB, Canton JH. 1993. Validation of some extrapolation methods with toxicity data derived from multiple species experiments. Ecotoxicol Environ Saf 25:341–359.

Okkerman PC, Van de Plassche EJ, Slooff W, Van Leeuwen CJ, Canton JH. 1991. Ecotoxicological effects assessment: a comparison of several extrapolation procedures. Ecotoxicol Environ Saf 21:182–191.

Oorts K, Ghesquiere U, Smolders E. 2007. Leaching and aging decrease nickel toxicity to soil microbial processes in soils freshly spiked with nickel chloride. Environ Toxicol Chem 26(6):1130–1138.

Oorts K, Ghesquiere U, Swinnen K, Smolders E. 2006a. Soil properties affecting the toxicity of $CuCl_2$ and $NiCl_2$ for soil microbial processes in freshly spiked soils. Environ Toxicol Chem 25:836–844.

Oorts K, Bronckaers H, Smolders E. 2006b. Discrepancy of the microbial response to elevated copper between freshly spiked and long-term contaminated soils. Environ Toxicol Chem 25(3):845–853.

Ovreas L, Torsvik VV. 1998. Microbial diversity and community structure in two different agricultural soil communities. Microbiol Ecol 36:303–315.

Parker DR, Pedler JF, Ahnstrom ZAS, Resketo M. 2001. Reevaluating the free-ion activity model of trace metal toxicity towards higher plants: experimental evidence with copper and zinc. Environ Toxicol Chem 20:899–906.

Pedersen F, Kristensen P, Damborg A, Christensen HW. 1994. Ecotoxicological evaluation of industrial wastewater. Miljøprojekt 254. Copenhagen (Denmark): Ministry of Environment.

Peijnenburg WJGM, Zablotskaja M, Vijver MG. 2007. Monitoring metals in terrestrial environments within a bioavailablity framework and a focus on soil extraction. Ecotox Environ Saf 67:163–179.

Pettit NM, Gregory LJ, Freedman RB, Burns RG. 1977. Differential stabilities of soil enzymes. Assay and properties of phosphatase and arylsulphatase. Acta Biochim Biophys 485:357–366.

Plackett RL, Hewlett PS. 1952. Quantal responses to mixtures of poisons. J R Stat Soc B 14:141–163.

Plette ACC, Nederlof MM, Temminghoff EJM, van Riemsdijk WH. 1999. Bioavailability of heavy metals in terrestrial and aquatic systems: a quantitative approach. Environ Toxicol Chem 18:1882–1890.

Posthuma L, Notenboom J. 1996. Toxic effects of heavy metals in three worm species exposed in artificially contaminated soil substrates and contaminated field soils. Report 719102048. Bilthoven (Netherlands): RIVM. 79 p.

Posthuma L, Suter GW, Traas T. 2001. Species sensitivity distributions in ecotoxicology. Boca Raton (FL): CRC Pr.

Rand GM, Wells PG, McCarty LS. 1995. Introduction to aquatic toxicology. In: Rand GM, editor. Fundamentals of aquatic toxicology. Effects, environmental fate and risk assessment. 2nd ed. Washington (DC): Taylor and Francis. p 367.

Reddy MR, Perkins HF. 1974. Fixation of zinc by clay minerals. Soil Sci Soc Am J 38:229–231.

Reimann C, Garrett RG. 2005. Geochemical background — concept and reality. Sci Total Environ 350(1–3):12–27.

Richards LA. 1954. Diagnosis and improvement of saline and alkali soils. Washington (DC): U.S. Salinity Laboratory, U.S. Dept. Agriculture.

Ritz K. 2007. The plate debate: cultivable communities have no utility in contemporary environmental microbial ecology. FEMS Microbiol Ecol 60:358–362.

Rooney CP, Zhao F-J, McGrath SP. 2006. Soil factors controlling the expression of copper toxicity to plants in a wide range of European soils. Environ Toxicol Chem 25:726–732.

Rooney CP, Zhao F-J, McGrath SP. 2007. Phytotoxicity of nickel in a range of European soils: influence of soil properties, Ni solubility and speciation. Environ Pollut 145:596–605.

Ross HLB. 1996. The interaction of chemical mixtures and their implications on water quality guidelines [Hons thesis]. Sydney, NSW (Australia): University of Technology. 167 p.

Ross HLB, Warne M. 1997. Most chemical mixtures have additive aquatic toxicity. Proceedings of the Third Annual Conference of the Australasian Society for Ecotoxicology; 1997 July 17–19; Brisbane, Queensland (Australia): 30 p.

Rufyikiri G, Genon JG, Dufey JE, Delvaux B. 2003. Competitive adsorption of hydrogen, calcium, potassium, magnesium, and aluminum on banana roots: experimental data and modeling. J Plant Nutr 26:351–368.

Rusk JA, Hamon RE, Stevens DP, McLaughlin MJ. 2004. Adaptation of soil biological nitrification to heavy metals. Environ Sci Technol 38:3092–3097.

Rutgers M, Sweegers BMC, Wind BS, van Veen RPM, Folkerts AJ, Posthuma L, Breure AM. 1998. Pollution-induced community tolerance in terrestrial microbial communities. In: Proceedings of the Sixth International FZK/TNO Conference on Contaminated Soil (Contaminated Soil '98). Edinburgh (UK). London: Thomas Telford. p 337–343.

Sample BE, Beauchamp JJ, Efroymson R, Suter GW. 1999. Literature-derived bioaccumulation models for earthworms: development and validation. Environ Toxicol Chem 18(9):2110–2120.

Sarkar D, Mandal B, Mazumdar D. 2008. Plant availability of boron in acid soils as assessed by different extractants. J Soil Sci Plant Nutr 171:249–254.

Schudoma D. 1994. Derivation of water quality objectives for hazardous substances to protect aquatic ecosystems: single-species test approach. Environ Toxicol Water Qual 9:263–272.

Schultz E, Joutti A, Räisänen M-L, Lintinen P, Martikainen E, Lehto O. 2004. Extractability of metals and ecotoxicity of soils from two old wood impregnation sites in Finland. Sci Tot Environ 326:71–84.

Schwertfeger D, Hendershot W. 2008. Differences in copper bioavailability and phytotoxicity in leached and non-leached spiked test soils. Paper presented at SETAC World Congress, 3–7 August 2008, Sydney (Australia).

Scott-Fordsmand JJ, Stevens D, McLaughlin M. 2004. Do earthworms mobilize fixed zinc from ingested soil? Environ Sci Technol 38:3036–3039.

Semenzin E, Temminghoff EJM, Marcomini A. 2007. Improving ecological risk assessment by including bioavailability into species sensitivity distributions: an example for plants exposed to nickel in soil. Environ Pollut 1488:642–647.

Shao QX. 2000. Estimation for hazardous concentrations based on NOEC toxicity data: an alternative approach. Environmetrics 11:583–595.

Shen GQ, Lu YT, Hong JB. 2006. Combined effect of heavy metals and polycyclic aromatic hydrocarbons on urease activity in soil. Ecotoxicol Environ Saf 63(3):474–480.

Shi Z, Di Toro DM, Allen HE, Sparks DL. 2008. A WHAM-based kinetics model for Zn adsorption and desorption to soils. Environ Sci Technol 42:5630–5636.

Sijm D, De Bruijn J, Crommentuijn T, Van Leeuwen K. 2001. Environmental quality standards: endpoints of triggers for a tiered ecological effect assessment approach? Environ Toxicol Chem 20(11):2644–2648.

Smit CE, Van Gestel CAM. 1998. Effects of soil type, prepercolation and aging on bioaccumulation and toxicity of zinc for springtail *Folsomia candida*. Environ Toxicol Chem 17:1132–1141.

Smith EP, Cairns J Jr. 1993. Extrapolation methods for setting ecological standards for water quality: statistical and ecological concerns. Ecotoxicology 2:203–219.

Smolders E. 2000. The effect of $NiSO_4 \times 6H_2O$, elemental Ni and green NiO on nitrogen transformation in soil. Final report for Nickel Producers Environmental Research Association, Durham (NC), Leuven (Belgium): Katholieke Universiteit Leuven. 22 p.

Smolders E, Brans K, Coppens F, Merckx R. 2001. Potential nitrification rate as a tool for screening toxicity in metal-contaminated soils. Environ Toxicol Chem 20:2469–2474.

Smolders E, Buekers J, Oliver I, McLaughlin MJ. 2004. Soil properties affecting toxicity of zinc to soil microbial properties in laboratory-spiked and field-contaminated soils. Environ Toxicol Chem 23:2633–2640.

Smolders E, Buekers J, Waegeneers N, Oliver I, McLaughlin MJ. 2003. Effects of field and laboratory Zn contamination on soil microbial processes and plant growth. Final report to the International Lead and Zinc Research Organisation. Leuven (Belgium): Katholieke Universiteit Leuven and CSIRO. 67 p.

Smolders E, Degryse F. 2002. Fate and effect of zinc from tire debris in soil. Environ Sci Technol 36:3706–10.

Smolders E, Lambregts RM, McLaughlin MJ, Tiller KG. 1998. Effect of soil solution chloride on cadmium availability to Swiss chard. J Environ Qual 27:426–431.

Smolders E, McLaughlin MJ. 1996. Chloride increases cadmium uptake in Swiss chard in a resin-buffered nutrient solution. Soil Sci Soc Am J 60:1443–1447.

Smolders E, Oorts K, Van Sprang P, Schoeters I, Janssen CJ, McGrath SP, McLaughlin MJ. 2009. Toxicity of trace metals in soil as affected by soil type and aging after contamination: using calibrated bioavailability models to set ecological soil standards. Environ Toxicol Chem 28:1633–1642.

Sneller FEC, van Heerwaarden LM, Schat H, Verkleij JAC. 2000. Toxicity, metal uptake, and accumulation of phytochelatins in *Silene vulgaris* exposed to mixtures of cadmium and arsenate. Environ Toxicol Chem 19(12):2982–2986.

Song J, Zhao F-J, Luo Y-M, McGrath SP, Zhang H. 2004. Copper uptake by *Elsholtzia splendens* and *Silene vulgaris* and assessment of copper phytoavailability in contaminated soils. Environ Pollut 128:307–315.

Song J, Zhao F-J, McGrath SP, Luo Y-M. 2006. Influence of soil properties and aging on arsenic phytotoxicity. Environ Toxicol Chem 25:1663–1670.

Speir TW, Kettles HA, Percival HJ, Parshotam A. 1999. Is soil acidification the cause of biochemical responses when soils are amended with heavy metal salts? Soil Biol Biochem 31:1953–1961.

Spurgeon DJ, Lofts S, Hankard PK, Toal M, McLellan D, Fishwick S, Svendsen C. 2006. Effect of pH on metal speciation and resulting metal uptake and toxicity for earthworms. Environ Toxicol Chem 25:788–796.

Steenbergen NTTM, Iaccino F, De Winkel M, Reijnders L, Peijnenburg WJGM. 2005. Development of a biotic ligand model and a regression model predicting acute copper toxicity to the earthworm *Apporectodea caliginosa*. Environ Sci Technol 39:5694–5702.

Stephan CE, Mount DI, Hansen DJ, Gentile JH, Chapman GA, Brungs WA. 1985. Guidelines for deriving numerical national water quality criteria for the protection of aquatic organisms and their uses. Report PB-85-227049. Washington (DC): USEPA. 98 p.

Stevens DP, McLaughlin MJ, Heinrich T. 2003. Determining toxicity of lead and zinc runoff in soils: salinity effects on metal partitioning and on phytotoxicity. Environ Toxicol Chem 22(12):3017–3024.

Struijs J, van de Meent D, Peijnenburg WJGM, van de Hoop MAGT, Crommentuijn T. 1997. Added risk approach to derive maximum permissible concentrations for heavy metals: how to take natural background levels into account. Ecotoxicol Environ Saf 37:112–118.

Taylor JP, Wilson M, Mills S, Burns RG. 2002. Comparison of microbial numbers and enzymatic activities in surface soils and subsoils using various techniques. Soil Biol Biochem 34:387–401.

Thakali S, Allen HE, Di Toro DM, Ponizovsky AA, Rooney CP, Zhao F-J, McGrath SP. 2006a. A terrestrial biotic ligand model. 1. Development and application to Cu and Ni toxicities to barley root elongation in soils. Environ Sci Technol 40:7085–7093.

Thakali S, Allen HE, Di Toro DM, Ponizovsky AA, Rooney CP, Zhao F-J, McGrath SP, Criel P, Van Eeckhout H, Janssen CR, Oorts K, Smolders E. 2006b. Terrestrial biotic ligand model. 2. Application to Ni and Cu toxicities to plants, invertebrates, and microbes in soil. Environ Sci Technol 40:7094–7100.

USEPA. 1996. Soil screening guidance: technical background. Report EPA/540/R-95/128. Washington (DC): USEPA.

USEPA. 2005. Guidance for developing ecological soil screening levels. OSWER Directive 9285.7-55. Washington (DC): Office of Solid Waste and Emergency Response, USEPA. 85 p. Also available at: http://www.epa.gov/ecotox/ecossl/.

USEPA. 2007a. Framework for metals risk assessment. Report 120/R-07/001. Washington (DC): USEPA.

USEPA. 2007b. Aquatic life ambient freshwater quality criteria—copper. 2007 Rev. Report EPA–822–R–07–001. Washington (DC): Office of Water, USEPA. 49 p.

Van de Meent D, Aldenberg T, Canton JH, van Gestel CAM, Slooff W. 1990. Background study for the policy document "Setting environmental quality standards for water and soil." Rapport 670101 002. Bilthoven (Netherlands): National Institute of Public Health and the Environment (RIVM).

Van de Plassche EJ, Polder MD, Canton JH. 1993. Derivation of maximum permissible concentrations for several volatile compounds for water and soil. Report 679101 008. Bilthoven (Netherlands): National Institute of Public Health and Environment Protection.

Van Gestel CAM, Koolhaas JE. 2004. Water-extractability, free ion activity, and pH explain cadmium sorption and toxicity to *Folsomia candida* (Collembola) in seven soil-pH combinations. Environ Toxicol Chem 23:1822–1833.

Vance ED, Brookes PC, Jenkinson DS. 1987. An extraction method for measuring soil microbial biomass C. Soil Biol Biochem 19:703–707.

Vandenhove H, Antunes K, Wannijn J, Duquene L, Van Hees M. 2007. Method of diffusive gradients in thin films (DGT) compared with other soil testing methods to predict uranium phytoavailability. Sci Tot Environ 373:542–555.

Van Leeuwen CJ, Bro Rasmussen F, Feijtel TCJ, Arndt R, Bussian BM, Calamari D, Glynn P, Grandy NJ, Hansen B, Van Hemmen JJ, Hurst P, King N, Koch R, Muller M, Solbe JF, Speijers GAB, Vermeire T. 1996. Risk assessment and management of new and existing chemicals. Environ Toxicol Phar 2:243–299.

Van Straalen NM, Denneman CAJ. 1989. Ecotoxicological evaluation of soil quality criteria. Ecotoxicol Environ Saf 18:241–251.

Vijver M, Jager T, Posthuma L, Peijnenburg W. 2001. Impact of metal pools and soil properties on metal accumulation in *Folsomia candida* (Collembola). Environ Toxicol Chem 20:712–720.

Vlaams Reglement Bodemsanering (VLAREBO). 2008. Flemish soil remediation decree ratified 14 December 2007. Published 22 April 2008.

Wagner C, Lokke H. 1991. Estimation of ecotoxicological protection levels from NOEC toxicity data. Water Res 25:1237–1242.

Wang H. 1987. Factors affecting metal toxicity to (and bioaccumulation by) aquatic organisms—overview. Environ Int 13:437–457.

Warne M. 1998. Critical review of methods to derive water quality guidelines for toxicants and a proposal for a new framework. Scientist Report 135. Canberra, ACT, (Australia): Supervising Scientist. 82 p.

Warne M, Hawker DW. 1995. The number of components in a mixture determines whether synergistic, antagonistic or additive toxicity predominate: the funnel hypothesis. Ecotoxicol Environ Saf 31:23–28.

Warne M, Heemsbergen DA, McLaughlin MJ, Kookana R. 2009. Proposed soil quality guidelines for arsenic, chromium (III), copper, DDT, lead, naphthalene, nickel and zinc. CSIRO Land and Water Science Report 44/09. Report prepared for the National Environment Protection Authority. Adelaide (Australia).

Warne M, Heemsbergen DA, McLaughlin MJ, Bell M, Broos K, Whatmuff M, Barry G, Nash D, Pritchard D, Penney N. 2008b. Models for the field-based toxicity of copper and zinc salts to wheat in eleven Australian soils and comparison to laboratory-based models. Environ Pollut 156(3):707–714.

Warne M, Heemsbergen D, Stevens D, McLaughlin M, Cozens G, Whatmuff M, Broos K, Barry G, Bell M, Nash D, Pritchard D, Penney N. 2008a. Modeling the toxicity of copper and zinc salts to wheat in 14 soils. Environ Toxicol Chem 27:786–792.

Warne M, Van Dam R. 2008. NOEC and LOEC data should no longer be generated or used. Aust J Ecotoxicol 14(1):1–5.

Wendling LA, Ma Y, Kirby JM, McLaughlin MJ. 2009. A predictive model of the effects of aging on cobalt fate and behaviour in soil. Environ Sci Technol 43(1):135–141

Weng LP, Temminghoff EJM, Lofts S, Tipping E, Van Riemsdijk WH. 2002. Complexation with dissolved organic matter and solubility control of heavy metals in a sandy soil. Environ Sci Technol 36:4804–4810.

Weng LP, Wolthoorn A, Lexmond TM, Temminghoff EJM. 2004. Understanding the effect of soil characteristics on phytotoxicity and bioavailability of nickel using speciation models. Environ Sci Technol 38:156–162.

Wheeler JR, Grist EPM, Leung KMY, Morritt D, Crane M. 2002. Species sensitivity distributions: data and model choice. Mar Poll Bull 45:192–202.

Widmer F, Shaffer BT, Porteous LA, Seidler RJ. 1999. Analysis of nifH gene pool complexity in soil and litter at a Douglas fir forest site in the Oregon cascade mountain range. Appl Environ Microbiol 65:374–380.

Wilke B, Mai M, Gattinger A, Schloter M, Gong P. 2005. Effects of fresh and aged copper contaminations on soil microorganisms. J Plant Nutr Soil Sci 168:668–675.

Witter E, Gong P, Baath E, Marstorp H. 2000. A study of the structure and metal tolerance of the soil microbial community six years after cessation of sewage sludge applications. Environ Toxicol Chem 19(8):1983–1991.

Yermiyahu U, Rytwo G, Brauer DK, Kinraide TB. 1997. Binding and electrostatic attraction of lanthanum (La^{3+}) and aluminum (Al^{3+}) to wheat root plasma membranes. J Membr Biol 159:239–252.

Zarcinas BA, Ishak CF, McLaughlin MJ, Cozens G. 2004. Heavy metals in soils and crops in Southeast Asia: 1. Peninsular Malaysia. Environ Geochem Health 26:343–357.

Zarcinas BA, Pongsakul P, McLaughlin MJ, Cozens G. 2003. Heavy metals in soils and crops in Southeast Asia: 2. Thailand. Environ Geochem Health 26(4):359–371.

Zelles L. 1996. Fatty acid patterns of microbial phospholipids and lipopolysaccharides. In: Schinner F, Öhlinger R, Kandeler E, Margesin R., editors. Methods in soil biology. Berlin: Springer. p 80–93.

Zhang H, Davison W. 1995. Performance-characteristics of diffusion gradients in thin-films for the in-situ measurement of trace-metals in aqueous-solution. Anal Chem 67:3391–3400.

Zhang H, Davison W, Gadi R, Kobayashi T. 1998. In situ measurement of dissolved phosphorus in natural waters using DGT. Anal Chim Acta 370:29–38.

Zhang H, Lombi E, Smolders E, McGrath S. 2004. Kinetics of Zn release in soils and prediction of Zn concentration in plants using diffusive gradients in thin films. Environ Sci Technol 38:3608–3613.

Zhang H, Zhao FJ, Sun B, Davison W, McGrath SP. 2001. A new method to measure effective soil solution concentration predicts copper availability to plants. Environ Sci Technol 35:2602–2607.

Zhao FJ, McGrath SP, Merrington G. 2007. Estimates of ambient background concentrations of trace metals in soils for risk assessment. Environ Pollut 148(1):221–229.

Zhao FJ, Rooney CP, Zhang H, McGrath SP. 2006. Comparison of soil solution speciation and diffusive gradients in thin-films measurement as an indicator of copper bioavailability to plants. Environ Toxicol Chem 25:733–742.

3 Variation in Soil Quality Criteria for Trace Elements to Protect Human Health

Exposure and Effects Estimation

Beverley Hale, Nick Basta, Craig Boreiko, Teresa Bowers, Betty Locey, Michael Moore, Marylène Moutier, Leonard Ritter, Erik Smolders, Ilse Schoeters, and Shu Tao

3.1 INTRODUCTION

The purpose of this chapter is to explore the variation among jurisdictions in soil quality standard (SQS) for the protection of human health, with the objective of understanding how these variations might be rationally harmonized. An SQS for the protection of human health typically varies with land use, as the relative contribution of each of the human exposure pathways (inhalation, soil ingestion, and food chain exposure) to trace elements (TEs) also varies with land use. While human health risk assessment is typically considered to be much simpler than ecological risk assessment because there is only 1 species to protect, there are various human health endpoints to be protected. For some TEs, there are important endpoints for each of the exposure pathways (inhalation, ingestion) so the SQS depends on the lowest exposure that reaches a critical value. Thus the endpoint and SQS can vary with land use, as this influences the type of exposure.

The authors of this chapter concluded that the variation among jurisdictions in their determination of "exposure" was greater than those for determination of "effects." Exposure and effects are the numerator and denominator, respectively, of a tolerable daily intake (TDI) or a reference dose (RfD). Most of this discussion focuses on estimates of exposure for TEs.

3.2 EXPOSURE CHARACTERIZATION

3.2.1 Background Exposure

Background exposure consists of all sources of human exposure to that TE in media other than those arising from the soil for which the threshold is derived. It includes inhalation of ambient air, ingestion of soils, dusts, and foods other than those originating from that soil, drinking water that is not groundwater from the contaminated site, and use of consumer products. For some TEs, background exposure might be mostly at concentrations that are preanthropogenic, but for other TEs, background exposure can be considerably elevated. The adjustment of SQSs for background exposure is intended to ensure that additional risk to human health from soil contaminants does not allow total exposure beyond that which is thought to cause adverse health effects. Not all jurisdictions correct SQSs for background exposure, and for those that do, the corrections are not made uniformly.

3.2.2 Comparison among Jurisdictions

"Threshold contaminants" are those for which there is no observed effect concentration (NOEC); i.e., there is a dose below which there is no predicted adverse effect on human health. For these contaminants, toxicity may typically occur over a narrow range of exposure doses above this threshold. A TDI for these contaminants is determined from the threshold dose, from which environmental criteria are back-calculated. For threshold contaminants only, some jurisdictions (i.e., Canada, United Kingdom, Belgium, and Germany) subtract background exposure from the TDI, leaving an residual tolerable daily intake (RTDI) attributable to the human health risk from TEs in contaminated soil only (Provoost et al. 2006). Sweden considers background exposure to Cd, Hg, Ni, and Pb in their derivation of soil guideline values for these elements, which are adjusted downward to reflect the substantial input expected from sources other than the contaminated soil (Elert et al. 1997). Similarly, the United States adjusts the TDI for Mn downward in its application to assessing human health risk from contaminated soils (in the IRIS database) to recognize the dietary contribution to total human exposure to this TE (US Environmental Protection Agency [USEPA] 2004). Other jurisdictions do not include such consideration of background exposure in the algorithms for the determination of SQSs (Netherlands, France, Norway, and Switzerland). Some of these values are based on ecotoxicity risk, which is lower than the SQS, to protect human health. For some TEs, background exposure can be a significant contribution to the total exposure, leaving a very small portion of the TDI as residual. This is not likely for many TEs but could be very significant for a few, particularly those where there is high natural background (such as As in drinking water) or elevated concentrations in foods because of limited physiological restrictions in soil-to-plant transfer (e.g., Cd and Mn) (Chaney 1980; Basta et al. 2005). The result of this approach is that the tolerance for additional exposure from contaminated soils for some TEs is very small, potentially leading to cleanup requirements for contaminated soils that are disproportionately punitive to some sources of exposure.

TABLE 3.1
Range in generic soil quality standard calculated

	Based on RTDI (i.e., background exposure considered) (mg kg^{-1})	Based on other criteria, including ecotoxicity (mg kg^{-1})
Arsenic	12–50, 110a	2–55
Cadmium	0.4–30	3.0–3.7
Copper	63, 400b	100–3100
Chromium III	64–400	25–100,000
Lead	80–700	60–1000
Mercury	1–20	1–23
Nickel	35–470	50–1600
Zinc	200, 1000b	100–23,000

[a] All values except 110 were between 12 and 50.
[b] Only 2 values.

As might be anticipated, the ranges of SQSs calculated as an RTDI are generally lower for many elements than are the SQSs calculated using other criteria (Table 3.1). However, the highest value for the "other criteria" SQS for each of Cu, Cr, Ni, and Zn (all are USEPA criteria) are quite out of the range of the remaining values for those elements, and if omitted, the range of remaining criteria values for each TE becomes much more similar to its RTDI counterpart.

The SQSs for the jurisdictions that do consider background exposure to TEs still can range 10-fold or more (Table 3.1), something that may be partially explained by the various assumptions and/or data used to estimate background exposure. The assumptions and uncertainties that go into the selection of the TDI value potentially have a significant impact on the size of the RTDI. These assumptions and uncertainties will not be discussed further here, beyond noting that identifying a TE as a threshold contaminant (thus having a TDI) can arise from both scientific and policy considerations. For example, As is now considered in some human health risk assessments a nonthreshold contaminant, although an "acceptable" dose still should be defined, i.e., one that results in acceptable risk of harm. It could be argued that to protect human health, SQSs should not be adjusted for background exposure, as any exposure will cause some degree of harm. Practically, however, this can result in an SQS that is lower than the local geogenic background and thus not achievable. Jurisdictions with elevated As in groundwater, for example, will accommodate this background exposure in the setting of SQSs. The USEPA now considers Pb to be a nonthreshold contaminant, as human health effects are predicted for children from any incremental increase in blood Pb.

3.2.3 Proportion of Total Exposure Allocated to Background

The first assumption to consider is the proportion of total exposure that is allocated to "background." Conceptually, land use largely determines the fraction of metal that affects the exposure. For example, if SQSs relate to urban soils, then TE exposure via the food chain can be estimated from the fraction food grown locally in the kitchen garden (Section 3.5.1). Pragmatically, regulators may select a given percentage of the TDI that is attributed to the contaminated land. In Canada, estimated daily intake (EDI) (i.e., background) is calculated for soil, consumer products, air, water, and food total exposure; these are totaled, subtracted from the TDI to give the RTDI. The RTDI is then divided evenly among the 5 exposure media, i.e., 20% of the RTDI is attributable to contaminated soil. The choice of 20% as the proportion is somewhat arbitrary and is an acknowledgement that it is very difficult to determine the true proportion contributed by each medium with any certainty. The generic SQS is then set considering the sum of the background exposure to soil (from the EDI) plus 20% of the RTDI. If the EDI is sufficiently high so that there is no RTDI (i.e., the EDI = TDI), then a provisional soil quality criterion is the background concentration in soil or the limit of quantification (Canadian Council of Ministers of the Environment [CCME] 2006).

In contrast, the United Kingdom (Contaminated Land Exposure Assessment [CLEA] 2008) sets the maximum proportion of TDI attributable to background exposure at 50%; thus the minimum exposure attributable to contaminated soil is 50% of total (the "50% rule"). In the case where there are separate TDIs for ingestion and inhalatory routes of exposure (as for Ni, for example), then the background exposure for each of these is calculated and separately compared to the relevant TDI. If there is only 1 TDI, then background exposures from all routes are summed and compared to the TDI unless there is compelling physiological or toxicological evidence of a barrier between 1 or more of the routes and the site of action (CLEA 2008). In the case where the same TE has both nonthreshold and threshold effects, background exposure for the index dose (ID) (i.e., nonthreshold) routes is not subtracted from the TDI routes of exposure. The aggregate exposure for all age classes included in the assessment is used to determine whether the default 50% rule is applied; for individual age classes, proportional exposure may vary. Then the sum of the background and contaminated soil exposure is used to determine the hazard quotient. In practice, for Cd, Ni, and Pb, CLEA estimates background exposure for the residential land use class to be very close to 50%. In Sweden, background exposure is considered in the setting of soil guidelines only for Cd, Hg, Ni, and Pb by attributing a fixed percentage of the TDI for each of these contaminants to background exposure (Cd, 25%; Hg, 70%; Ni, 50%; Pb, 33%), leaving the remainder for attribution to contaminated soil (Elert et al. 1997).

Not only do jurisdictions vary the percentage of total exposure attributable to background, but few of these schemes for attribution of total exposure are likely to accurately reflect actual background exposure for many age groups. For example, dietary exposure to Cd, alone, for a number of population age groups, can account for upward of 90% of the TDI (Olsson et al. 2005) (Section 3.5). These calculations do not include exposure to Cd through smoking, which is likely to overwhelm all other sources of exposure (particularly because pulmonary absorption is typically much greater than gastrointestinal). It can also be a significant source of exposure for nonsmokers via residential particulate. In the determination of the SQS for Cd,

TABLE 3.2
Soil quality standards for Canada, Sweden, and the United Kingdom for trace elements in residential soil[a]

	Cadmium	Nickel	Lead
Canada	10	50	140
Sweden	0.4	35	80
United Kingdom	10	130	450

[a] Trace elements are adjusted for background exposure (mg kg^{-1}).

this self-selected population is not typically considered a "sensitive" group that must be protected.

Differences among jurisdictions in what proportion of total background exposure is attributed to each of the pathways that contribute to background do not explain the variation among SQSs for a TE (Table 3.2). For example, the UK SQSs for Ni and Pb is higher than the comparable Canadian value, roughly consistent with the difference between 20% of the RTDI plus the soil EDI (CCME 2006) and not less than 50% of the TDI (CLEA 2008). Following the same logic, the SQS for Cd in Canada should be higher than that for the United Kingdom; however, these values are the same. In the UK SQS, consumption of backyard produce and associated soil particles is the largest contributor to exposure for residential land use. However, Sweden's background-adjusted SQS for Cd (0.4 mg kg^{-1} for less sensitive land considering groundwater extraction) does not fit this continuum at all, as it would be expected to be consistent with 75% of the TDI (Elert et al. 1997). The reason for this discrepancy is that Sweden's SQS for Cd is based on the dominant exposure pathway being ingestion of groundwater as drinking water and includes consumption of garden produce in the background exposure. The same rationale is applied to Sweden's SQSs for Ni and Pb; thus they are each the lowest of the 3 countries.

3.3 INHALATION

On an average working day, men and women with sedentary occupations will inhale about 23 and 19 m^3 air, respectively (Roy and Courtay 1991). Adults engaged in heavy manual work will inhale at least 15% more air, and children will inhale less than that of typical adults. Some jurisdictions make adjustments of inhalation rates for age (e.g., CLEA 2008), and still others predict entrainment of dust particles into the air for different land use scenarios (Elert et al. 1997). Each cubic meter of outdoor air holds billions of suspended particles arising from both pollution and natural sources such as soils (Holmes et al. 2005). Inhaling these particles is unavoidable, and many will be deposited somewhere in the respiratory tract or in the lungs. With respect to human exposure to soils and dusts, it is acknowledged that inhalation of soils and dusts with subsequent pulmonary absorption of TEs will normally be a lesser source of those elements than alimentary and gastrointestinal exposure and absorption. It is also clear that some particulate material especially in the larger size range, initially taken into the upper airways, will be cleared by ciliary action and pass through the esophagus into the gut. The primary determinant of pulmonary adsorption is that the particle size must be sufficiently small for the particles to pass into the lower airways, the alveolar space, and then into the circulation. The ability therefore to consider inhalation as a pathway to include in the determination of SQSs will be predicated on measurement of the size distribution of particulate material generated by the soil under examination and on the likelihood that it is entrained or reentrained into the air column associated with human exposure.

3.3.1 Particle Size Domain

Conventionally, particulate size is inferred from measurement of weight, the particle matter (PM) standards (i.e., 150 μg m^{-3}, 24-hour standard for PM_{10}) (USEPA 2008). In human exposure these are normally linked to PM_{10} and $PM_{2.5}$ standards corresponding to less than 10- and 2.5-μm particulates. More recently, it has been recognized that smaller particulates (ultrafine particulates [UFP] or nanoparticulates), i.e., less than 0.1 μm, have a much greater capacity to penetrate deep lung space and hence to have physiological and toxicological impacts. This is due to their small size, which allows transmembrane passage, and to the proportionally greater surface area per unit weight, which allows greater solubilization of their elemental content (Morawska et al. 2004), and to the potential to generate adverse biological effects in living cells of a type that would not be possible with the same material in a larger bulk form (Nel et al. 2006). The main feature of the UFP size domain is that it occupies the transition zone between that of individual atoms or molecules and that of bulk materials (Nel et al. 2006). For example, a UFP particle with surface bound iron atoms can cross cellular membranes, presenting a persistent catalytic surface for generating reactive oxygen species into the mitochondria (Li et al. 2003a) in a way that is capable of overwhelming the physiological defenses against oxidative stress (Ghio and Cohen 2005).

Inhalation is the major route of exposure to UFP (Oberdörster et al. 2005). According to the International Commission on Radiological Protection (ICRP) 66 model, inhaled particles with the highest probabilities of being deposited in the most vulnerable region of the lung (namely, the respiratory bronchioles and the alveoli) are in the UFP size range (ICRP 1994). Particles around 12 nm have the highest probability of deposition in the alveoli; these particles are just 1/10th of the diameter of a laboratory replicated influenza A virus particle (Elleman and Barclay 2004). Particles in the UFP size domain represent by far the largest number of all particles deposited in the airway and/or lung following inhalation, but UFPs constitute an almost negligible fraction of the actual dose (or weight) of inhaled particulate material. Hygroscopic and readily water-soluble components of UFP will dissolve and lose their UFP identity on contact with the liquid lining of the airways and alveoli.

3.3.2 Particle Deposition

In humans the number of alveoli is complete by age 6, and subsequent lung volume increase is through increases in the dimensions of the airways and the alveoli

(Zeman and Bennett 2006). Particle deposition in the lungs is therefore a function of particle size and age of the individual. This is reflected in most deposition models (Asgharian et al. 2004). However, Massaro and DeCarlo Massaro (2007) argue that in the absence of environmental insults, alveolarization in humans might continue until cessation of somatic growth, as observed for Rhesus monkeys (Hyde et al. 2007). Under such a scenario, adults who experience respiratory insults during childhood might have a smaller reserve of alveoli to sustain gas exchange and lung recoil. Particle deposition and gas transport models generally assume that a transition between convection and diffusion occurs at airway generation 17 in adults (Sapoval et al. 2002). Three-dimensional computational fluid dynamic simulations conducted by Sznitman et al. (2007), which allowed for expansion and contraction of the alveoli, showed recirculating gas flow patterns in the proximal alveoli. However, for alveoli in the deep acinar generations the simulations showed largely radial flow patterns, implying convective transport to these alveoli.

The airways behave like a set of probabilistic filters connected in series. Particles bigger than 7 μm deposit with high probability in the region that includes the mouth, nose, and bronchi, and these particles are removed relatively quickly from the lower respiratory track by the mucociliary escalator. There is little probability of a 7-μm particle being deposited in the alveoli. Fine particulates deposit in all regions of the respiratory tract and in the alveoli. Gastrointestinal (GI) absorption is subject to substantial variation linked to the presence or absence of food in the GI tract. Unlike the gut, respiratory absorption is not modified by satiety and will therefore be more constant over time. If one deals with the particles on a count basis rather than a weight basis, the greater proportion will by far lie below 7 μm and, consequently, access the alveolar lung space. This will apply to all air-breathing species. A neglected aspect of particulate exposure is that the closer to the ground, the greater the likelihood of exposure. The hierarchy of exposure is companion animals > children > adults. The companion animals are thus potential sentinels for human exposure. However, to understand the toxicology in site-specific circumstances, there is ultimately a need to characterize the airborne soil-derived contribution. This will be best achieved through time-integrated passive sampling or other methods that establish the weight, count, size, and nature of the soil particles. It would be naïve to suppose that equal masses of soils and dust particulates would pass into the GI tract and be deposited in the lungs. It is generally assumed that around 100 mg of soil in all forms is ingested daily by adults with greater masses being ingested by children. No equivalent information is available for pulmonary intake largely because such inhalational intake has been discounted as being of no importance despite its centrality in the debate around abatement of lead pollution. Critically, the bioavailability of TEs is usually low for GI absorption but much higher for pulmonary absorption. For lead, the figures often used are 1 and 50%, respectively. If one takes the information given here and arbitrarily assigns 100 mg of intake to the gut and 2 mg to the lungs with 50% of respired intake returning to the gut (a value that is wholly contingent on the proportion of particles less than 7 μm), the net contribution to soil based elemental retention will then be 1 mg to the GI tract and 0.5 mg to the lungs. This demonstrates that the consequences of soil inhalation may not be trivial.

Observations by Kim and Jaques (2004, 2005) are consistent with theory concerning UFP deposition in the size range 40 to 100 nm. In particular, 1) diffusion processes dominate in the deposition of UFP in the lower respiratory tract and lung; 2) the implied diffusion coefficient increases with decreasing particle size in this size range, i.e., the smaller the particle size, the greater the deposition as a fraction of the count of particles inhaled; and 3) an increase in ventilation rate and mean respiratory time increases the deposition as a fraction of the count of particles inhaled. Kim and Jaques (2004, 2005) found no important differences between deposition of UFP in young and elderly adult subjects. For natural breathing patterns, UFP deposition was also found to increase as UFP particle size decreased (Daigle et al. 2003; Chalupa et al. 2004). Daigle et al. (2003) found that UFP deposition during exercise is higher than predicted by commonly applied deposition models. There is thus some continuing uncertainty about total or aggregate UFP deposition over the lifetime of an individual.

The model of Strum et al. (2006) illustrates pathways and tissue compartments involved in the translocation of UFP deposited in the lower respiratory tract and lung. The following important features have been consistently supported by experimental and epidemiological evidence:

1) The only pathways capable of eliminating of UFP deposited in the lower airway and lung from the body in the short term are those that deliver particles to the gastrointestinal tract.
2) The only UFPs that are eliminated relatively quickly (on the order of 48 hours) are those UFPs that deposit and remain in the mucociliary escalator until they are released to the gastrointestinal tract.
3) Clearance rates for the bronchopulmonary lymph nodes and the extracellular space of the lung interstitium may reasonably be taken to be so slow that these compartments can be regarded as UFP reservoirs.

3.3.3 Toxicity

All contemporary hypotheses concerning the pathways by which inhaled particulate matter produce acute toxics effects in humans rest on 1 of the following 2 principles:

1) Particles deposited in the airways or lung provoke a local inflammatory response that initiates inappropriate intercellular signaling that cascades into more serious harm.
2) Particles stimulate airway nerves resulting in an adverse effect on autonomic control of the heart.

An important issue for assessing the toxicity of UFP is whether UFP can cross the alveolar gas-blood barrier of the intact human epithelium and thereby translocate to the systemic circulation at a biologically significant rate. Rodent experiments have found that UFP deposited in the airways and/or lung 1) translocate to the blood vessels of the pulmonary circulation with the likelihood that UFP would also be found in the spleen, liver, and other organs (Geiser et al. 2005) or 2) translocate to the systemic circulation (Nemmar et al. 2002). Particle deposition models have generally ignored the epithelial lining fluid and surfactant. Scarpelli (2003) argued that epithelial lining fluid circulates in the liquid phase of a relatively dry foam that occupies the acini. The vast surface area of the foam would present minimal resistance to gas diffusion but would have a high resistance to particle diffusion in the gas phase.

On the basis of the clinical trials it seems likely that humans can cope following an acute exposure to environmentally relevant concentrations of UFPs that have no significant potential to establish intracellular redox cycling or alter intracellular calcium concentrations. However, environmental UFPs are not of this type. In particular, urban UFPs are associated with significant quantities of redox-cycling chemical species, i.e., iron, other transition elements, and quinones. Recent human clinical trials should not be viewed as being contrary to principles 1 or 2 stated earlier in this section. The correct interpretation of the trials is as a narrowing of the uncertainties of the mechanisms of action of environmental UFPs. In vitro and animal experiments suggest that inhaled UFPs have the capacity to translocate to the central nervous system (CNS) by adhesion and diffusion type processes in a nonmembrane bound state.

3.4 SOIL INGESTION

The soil ingestion pathway is often the most important route of exposure in determining SQSs, in particular for young children in a residential setting, and the soil ingestion rate is the most sensitive parameter in the soil ingestion exposure pathway. Therefore there is a high priority on reducing uncertainty in soil ingestion rates to the maximal extent possible. Compared to older children and adults, young children are generally expected to have the highest soil ingestion rates as a result of their higher hand-to-mouth behavior. However, there may be special circumstances where high soil ingestion rates are found in other receptors such as adults engaged in outdoor soil-intensive activities.

It is important to note that at present no distinction is made between ingestion of outdoor soil and ingestion of indoor dust, which includes a component of outdoor soil. The design of soil ingestion studies conducted to date does not allow this distinction to be made. This is an area where future research is needed. In addition, soil ingestion for adults, in particular, for adults with intensive soil contact activities, is also an area where further research is needed.

Van Holderbeke et al. (2007) provide a comprehensive review of the existing literature summarizing studies of soil ingestion rates in both children and adults. This review was developed to support risk assessment decision making at the Kempen area located at the border between Belgium (Flanders) and the Netherlands (and referred to as BeNeKempen project). The review considered several approaches to developing soil ingestion estimates, including 1) tracer studies in child and adult populations; 2) data on hand loading, soil-dust transfer efficiencies, and frequency of hand-to-mouth contact; 3) validation studies in which modeled and measured exposures are compared; and 4) analysis of empirical relationships between biomonitoring and environmental concentrations.

Van Holderbeke et al. (2007) concluded that the tracer studies were the most reliable means of estimating soil ingestion rates but noted that the other approaches provided important confirmation in yielding approximate soil ingestion rates consistent with those developed through the tracer studies. However, not all tracer studies were considered (e.g., data on Ti were left out because that tracer element was considered to be less reliable). To make their recommendations, Van Holderbeke et al. (2007) relied primarily on 1) results from tracer studies that correct the soil ingestion estimates to account for the potential contribution of tracer intake from other exposure routes than soil ingestion and 2) the most recent results when reanalyzed by the authors of the study.

Three tracer studies have been performed, while several analyses and reanalyses of the data presented in those studies have been subsequently published. Calabrese et al. (1989) studied 64 children between the ages of 1 and 4 from Massachusetts; Davis et al. (1990) studied 104 children between the ages of 2 and 7 from the state of Washington; Stanek and Calabrese (2000) studied 64 children between the ages of 1 and 4 years, residing at a Superfund site in Anaconda, Montana. These studies estimate soil ingestion by comparing the concentration of various tracer metals in soil and dust to the quantity of these tracers recovered from feces. By measuring the quantity of such tracers recovered from feces and then subtracting the amount of tracer that can be attributed to the ingestion of alternative materials, investigators can estimate the amount of soil and dust a child must have ingested.

Reanalyses of the original study data have centered on a methodology introduced by Stanek and Calabrese (1995) and referred to as the Best Tracer Method (BTM). The BTM restricts attention to measurements based on the 4 best tracers for each child, where a tracer's quality is ranked according to the ratio of total tracer in feces to total tracer in food (high ratios are best). The rationale for this ranking scheme is that measurement of tracer quantities in food is a source of so-called "framing error," which reflects the tendency to incorrectly match fecal tracer measurements with the temporally corresponding food tracer measurement. Tracers that are least present in food (i.e., tracers with high total tracer to food tracer quantity ratios) have the lowest potential for these errors, hence minimizing the influence of framing errors on the estimate of soil ingestion. The median of the 4 best tracers (calculated as the average of the results inferred using the second and third tracers ranked by estimated soil ingestion rate) is selected because individual tracers are sometimes subject to source attribution error. This type of error occurs when a tracer is unknowingly ingested via some medium other than food or soil (e.g., inadvertent ingestion of the tracer in toothpaste). Because source attribution errors tend to affect individual tracers, taking the median of several "good" tracers decreases the probability of this type of error.

Stanek and Calabrese (1995) published a reanalysis of their Amherst study data as well as the Washington State data published by Davis et al. (1990) using the BTM. The revised analysis calculated an average soil ingestion rate for each child over the period of the study. These average soil ingestion rates for each child then form a distribution of soil ingestion rates, which is approximately lognormal. Stanek and Calabrese (1995) report soil ingestion rates at percentiles of this distribution, e.g., for the 50th percentile child (50% of children have an average soil ingestion rate below this value) and the 95th percentile child (95% of children have an average soil ingestion rate below this value). The revised analysis suggested that the average soil ingestion rate for the 50th percentile child was 33 mg d^{-1} for the Amherst population, 44 mg d^{-1} for the population in the Davis et al. (1990) study, and 37 mg d^{-1} for the combined populations of the 2 studies. Average soil ingestion rates for the 95th percentile child were 154, 246, and 217 mg d^{-1} for the Amherst, Davis et al. (1990), and combined populations, respectively.

Stanek and Calabrese (2000) used the same approach described here in analyzing data from their Anaconda, Montana study, deriving a 7-day average soil ingestion rate for the 50th percentile child of 17 mg d^{-1}. The 7-day average soil ingestion rate for the 95th percentile child was 141 mg d^{-1}. Stanek and Calabrese (2000) also estimated average soil ingestion rates by extrapolating from the 7-day study period to longer periods. They estimated that the 95th percentile child will have a 365-day average soil ingestion rate of 106 mg d^{-1} for the Anaconda population and 124 mg d^{-1} for the Amherst population. These estimates are based on an analysis of uncertainty in the daily soil ingestion estimates, using standard statistical techniques. The estimates do not include an adjustment for seasonal effects (e.g., amount of frozen ground or snow cover in winter), which might be thought to further reduce yearlong average soil ingestion estimates below those measured in the 7-day summer studies for populations living in areas where grounds are frozen during winter. Stanek and Calabrese (2000, p. 634) do not present comparable long-term averages for the 50th percentile child.

There are major differences in soil ingestion rate estimates for the Amherst and Anaconda populations, whereas the population studied by Davis et al. (1990) appears to be similar to the Amherst population. It is unknown whether the differences in soil ingestion rates between the Amherst and Anaconda populations represent differences in climate, soil-to-dust transfer, knowledge among the Anaconda population that they are living on a Superfund site, or the "improved study design" (Stanek and Calabrese 2000, p. 634) of the Anaconda study. Differences in both population and study analysis may also play a role.

Van Holderbeke et al. (2007) summarize the results of these studies and estimate the 95th percentile upper confidence limits on median and mean estimates of daily soil ingestion rates. The authors present statistics for the 95th percentile child but do not recommend soil ingestion estimates at the upper end of the distribution. Van Holderbeke et al. (2007) also summarize the results of the adult soil ingestion studies. Table 3.3 summarizes our recommendations based on a combination of the recommendations made by Van Holderbeke et al. (2007) and the further information presented above concerning soil ingestion estimates at the upper percentiles of the population for children.

3.4.1 Recommended Soil Ingestion Values for Children Based on Tracer Studies

The mean daily soil intake rates from the literature examined by Van Holderbeke et al. (2007) range from 31 to 120 mg d^{-1} and the median daily values range from 17 to 42 mg d^{-1}. The authors used bootstrapping to generate additional data in order to gain more statistical information on the population distribution they come from.

TABLE 3.3
Distribution around the 95th percentile of the mean and median literature daily soil intake values

Type of daily soil ingestion literature data used[a]	Range of literature values	Distribution around arithmetic mean (mg d^{-1})				Distribution around the 95th percentile (mg d^{-1})		
		Arithmetic mean	50th Percentile	95th Percentile	Proposed indoor-outdoor soil ingestion value[a]	Arithmetic mean	50th Percentile	95th Percentile
Children (1 to 7 years)								
Mean values	3–120	63	63	81	60 (40–80)[b]	97	97	124
Median values	17–42	27	27	37	30 (20–40)[b]	38	37	50
95th percentile values	106–283	195	–	–		–	–	–
Adults								
Mean values	5–92	46	46	60	45	81	81	104
Median values	0.2–65[c]	24	24	41	25	56	55	81
75th percentile values[d]	37–120	65	–	–		–	–	–

Notes: The proposed values are rounded figures of the arithmetic mean of mean values or median values from the literature.

[a] Based on Van Holderbeke et al.'s (2007) recommendations.

[b] Corresponding to the 5th and 95th percentile range.

[c] Taking into account background exposure.

[d] The authors use the 75th percentile for the estimation of the upper bound soil ingestion rate based on Calabrese's (2003) recommendations.

The distribution around the arithmetic mean of the mean daily soil intake literature values results in a mean soil ingestion value of 63 mg d^{-1} (95th percentile = 81 mg d^{-1}). Rounded off, the proposed soil ingestion rate based on mean values from literature equals 60 mg d^{-1} (5th and 95th percentiles = 40 and 80 mg kg^{-1}, respectively).

The distribution around the arithmetic mean of the median daily soil intake literature values results in a mean soil ingestion value of 27 mg d^{-1} (95th percentile = 37 mg d^{-1}). Rounded off, the proposed soil ingestion rate based on median values from the literature equals 30 mg d^{-1} (5th and 95th percentiles = 20 and 40 mg d^{-1}, respectively).

Additional statistics are provided by Van Holderbeke et al. (2007) for the distribution around the 95th percentile of the mean and median literature daily soil intake values (Table 3.3). It should be noted that the USEPA child-specific exposure factors handbook (USEPA 2006) recommends a mean soil ingestion rate of 90 mg kg^{-1}, comparable with the arithmetic mean of the distribution around the 95th percentile of mean literature values.

3.4.2 Recommended Soil Ingestion Values for Adults Based on Tracer Studies

Although data regarding daily soil ingestion rates for adults are scarce, the literature review of Van Holderbeke et al. (2007) provides some statistics. The mean daily soil intake rates range between 5 and 92 mg d^{-1}, and median values range between 0.2 and 65 mg d^{-1}, taking into account the background exposure. More statistical information on the population distribution was obtained by applying the bootstrapping method.

The distribution around the arithmetic mean of the mean literature values results in a mean soil ingestion value of 46 mg d^{-1} (rounded to 45 mg d^{-1}) (95th percentile = 60 mg d^{-1}). The distribution around the 95th percentile has a mean value of 81 mg kg^{-1} (95th percentile = 104 mg kg^{-1}).

The distribution around the arithmetic mean of the median literature values results in a mean soil ingestion values of 24 mg d^{-1} (rounded to 25 mg d^{-1}) (95th percentile = 41 mg d^{-1}). The distribution around the 95th percentile of the median values has a mean value of 56 mg kg^{-1} (95th percentile = 81 mg kg^{-1}).

We recommend that soil ingestion estimates for both the 50th and 95th percentiles of the population be considered for the development of SQSs. Both can be used to form the basis for a probabilistic distribution of soil ingestion estimates for use in a probabilistic exposure assessment. Deterministic assessments should use the median soil ingestion estimates for a central estimate of risk and the 95th percentile soil ingestion estimates for a reasonable maximum exposure calculation.

Van Holderbeke et al. (2007) concluded that median and mean adult soil ingestion rates do not differ substantially from soil ingestion rates estimated for young children. This conclusion runs counter to the assumption typically made in many risk assessments that adult soil ingestion rates are approximately half, or less, those of children. The analysis presented by Van Holderbeke et al. (2007) may reflect a data gap in that adult soil ingestion rates have not been adequately studied. We recommend using Van

Holderbeke's recommendations, but further work is required to confirm their suitability. Soil ingestion rates presented here are based on populations in typical residential settings. All 3 tracer studies were performed in the United States. Although these studies likely have broad applicability, there may be countries or regions for which use of these recommendations are not appropriate because of conditions that would be expected to affect soil ingestion. For example, populations who live in adobe or dirt houses, as occurs in some regions, may have soil ingestion rates that depart substantially from those recommended here. Caution must be exercised in such situations, and uncertainties should be discussed in any application of these recommendations.

Additionally, the soil ingestion rates presented here are most applicable to residential exposure scenarios and may not reflect expected soil ingestion rates under scenarios that involve more intensive soil contact such as agricultural workers or adults engaged in other soil intensive contact work. Studies do not exist at the time to define soil ingestion rates expected under these alternate scenarios. We do expect, however, that the soil ingestion rates for young children presented here will be applicable in recreational scenarios as young children are typically engaged in soil contact activities in either their own yards or in recreational parks. Therefore SQSs developed on the basis of childhood exposure in a residential setting are applicable to a childhood recreational scenario as well.

3.5 FOOD CHAIN EXPOSURE

The food chain is 1 of the pathways by which humans are exposed to soil-borne TEs, and it includes 3 vectors: soil-plant-human, soil-plant-animal-human, and soil-animal-human. The latter 2 pathways result from both soil and crop-plant ingestion by animals. The extent by which an SQS depends on food chain transfer logically depends on the fraction of food produced on the land for which the SQS is developed. For example, it is logical that the SQS for agricultural soils highly depends on this pathway, whereas residential soils only depend on these in a kitchen garden scenario, unless background exposure to TEs via diet is included in exposure estimates for residential soils. The implementation of food chain transfer in an SQS typically starts with a decision on a land use scenario and the corresponding parameters for the fraction of food produced and, to gain precision, the type of food produced. The practical implementation of an SQS must also consider that land use can change over time.

A human health–based SQS derivation for agricultural land is sometimes simplified to the calculation of the concentration of a TE at which a legally enforced food limit is reached, e.g., the Codex TE limit for Cd in wheat grain. While this approach can be a pragmatic one, the SQS derived should not be termed a human health based value, rather it is a value that may be protective for marketing agricultural products. To be accurate, the various routes of the food chain transfer must be aggregated, and a collective flux of soil to human must be estimated. Each route is always the product of a concentration in a food item and the dietary intake of that food item, meaning that food chain calculation requires data on both dietary habits and on TE concentration in the food items and how these are affected by soil contamination. In the following, this approach is elaborated on by a discussion of land use scenarios, estimating dietary habits and estimating food chain contamination. Before doing that, we recommend that the large effort required to estimate food chain transfer should only be made if food chain exposure readily exceeds the TDI, or RTDI in the case of those jurisdictions that subtract background exposure. Data about background exposure are useful here, for example, showing that for Cd and As this transfer is highly important (more than 20% for average intake), while for Pb it is less important (less than 20% for average intake).

3.5.1 Land Use Scenarios: The Issue of Selecting Appropriate Background Exposure

The selection of land use scenarios is obviously a regulatory choice and here only technical details are discussed about the parameter values. In an agricultural scenario, food chain transfer is a major exposure pathway for TEs. The size of the area for which the SQS is derived can influence the ratio of background exposure to food chain exposure, although in some jurisdictions, this is a fixed ratio. This means that the fraction of an individual diet that is produced within the area must first be estimated. Here an implementation issue arises about the degree of food mixing within that area. An SQS for agricultural scenarios can obviously be applied to assess risk of an individual parcel; however, it is highly unlikely that an individual person is fully exposed to the TEs in that particular parcel of land during their lifetime. The question here is whether regulators are willing to accept the potential dilution aspect, i.e., that the food chain exposure from a parcel is diluted with that from other, less contaminated parcels within the same area. This issue is highly important if the SQS for an agricultural soil should be valid for a large agricultural area in which all sorts of food items are produced, including vegetables, cereals, and animal products, specifically, for situations where the background exposure (i.e., exposure unrelated to the area for which the SQS is valid) is relatively small. A conservative regulatory choice could be made to ignore the dilution, basing the SQS on the tolerable intake, i.e., the RTDI (TDI minus the very small background). For the residential scenarios the situation is easier to grasp and kitchen garden scenarios can be applied, e.g., assuming that 50% of food items are produced in that garden, the remainder considered as background. It is logical, then, that the conservative estimate for agricultural areas, i.e., assuming no dilution processes within the area, leads to significantly lower SQS values than corresponding values for kitchen garden scenarios.

3.5.2 Dietary Preferences

Data on dietary preferences are important to estimate the fluxes (the product of the mass of foods consumed and their TE concentrations) of TEs to humans through diet. Elevated concentrations of TEs in selected food items are often confused with elevated risk; however, that is often not the case when the dietary habits are taken into account. For example, dietary intake values at background for Cd show that the majority of the intake is through consumption of low concentration (on a dry weight basis) products (potatoes and grain) that are intensively consumed, while leafy vegetables with more elevated Cd hardly contribute owing to low consumption on a dry weight basis.

Dietary preferences vary among socioeconomic groups, with age, within populations, and among seasons. To protect the highly exposed individual (HEI), it is logical to assess the upper percentiles of dietary intake. The HEI is a person who has the largest estimated TE intake and is best estimated from the upper percentiles of TE intake values based on duplicate meals, with due accounting for seasonal trends. Some regulatory bodies have a detailed set of market basket data on TE concentrations in food items and consumption patterns of a sample of the population. Three remarks should be made to avoid overestimating exposure of the HEI. First, seasonal or more general day-to-day variation can be ignored because TDIs are generally to be used for lifetime exposure (unless otherwise stated in SQS guidance documents). This means than daily intake values should be based on data valid for lifetime average consumption of an individual person or, pragmatically, statistics based on several sampling days during the year; several regulatory bodies have such data. Second, the accumulation of worst case estimates by summing upper percentiles of each food item is a common error, as the person who consumes each food at the highest percentile intake does not likely exist. Finally, effects of age on the intake might be included. It is often observed that children, who have the largest dietary intake per unit of body weight, are the group where the HEI is found because the intake exceeds the body weight-based TDI most readily. Here again, the relevant guidance document for the TDI must be consulted: if lifetime exposure is important, then no body weight correction should be used and the HEI will be found among adults. In contrast, if shorter-term exposure matters and if children are inherently more sensitive, then the body correction should be applied and the HEI will be found within young age groups.

3.5.3 Soil-Plant Transfer

Risk assessment models are traditionally based on a first-order approach, i.e., the exposure is proportional to the environmental contamination, in each pathway. Soil-plant transfer is then quantified as crop/soil concentration ratios termed transfer factors (TFs) or bioaccumulation factors (BAFs). Experimental data confirm this concept; i.e., crop contamination rises about proportionally with increasing soil contamination provided that the concentration remains low and that the TE is added in a soluble form (e.g., Brown et al. 1998). Under these strict conditions, a BAF or TF is constant for a given soil and crop and can be used to derive an SQS. At higher concentrations, uptake becomes saturated because the ion uptake capacity of the plant is saturated, or toxicity limits accumulation, or elevated soil TE concentration consists of less bioavailable forms of the TE; consequently, the BAF or TF decreases. This nonlinearity is critical to consider when estimating exposure at sites with high TE concentrations in soil by using BAF or TF data derived from soils with low TE concentrations. Even within the linear phase of the relationship, field data show that concentrations of some TEs in above-ground plant parts exhibit a low correlation with the total concentration of the TE in soil. For example, agronomic large-scale surveys in wheat grain in the United Kingdom showed that less than 50% of the variation of crop Cd concentrations was explained by soil total Cd concentrations, while for Pb there was no significant association with soil concentration (Adams et al. 2004; Zhao et al. 2004). This lack of correlation is at odds with the concept in risk assessment assuming that exposure to TEs is related to soil contamination. The poor correlation is because of differences in soil TE availability, including competition for uptake between TEs at high concentrations with similar properties, the influence of meteorology on transport of TEs from roots to shoots, atmospheric deposition of TEs to above-ground plant parts, as well as homeostatic control of TEs that are essential for plants.

Difference in soil TE bioavailability is a factor that explains why surveys often fail to find a strong association between soil and crop contamination. Soils differ in TE availability because of differences in speciation of the element in soil that control the supply of the TE and because soil properties also affect the physiology of TE uptake (Adams et al. 2004) (Chapter 2). There are various possibilities to reduce the uncertainty: the TE availability can be predicted with soil tests (e.g., soil extraction) (McLaughlin et al. 2000) or with empirical regression models (e.g., Eriksson et al. 1996; Efroymson et al. 2001; Nan et al. 2002; Li et al. 2003b; McLaughlin et al. 2006). The former approach (soil extraction) is most attractive from a scientific point of view; however, the consequence is that SQSs are then also expressed on a soil extract basis rather than on total soil TE concentration. The disadvantage of this is that such an approach cannot be merged with the calculation of exposure via other pathways (soil ingestion) that are based on total soil concentration or another extraction method, i.e., a risk assessment with a complete pathway analysis is compromised when the pathways have no common basis for expressing the soil contamination. The second approach is mainly empirical in which the food chain exposure is empirically related to soil contamination using a BAF or TF that is a function of soil properties such as pH and texture. Here soil properties are required to predict the SQSs. For example, this has been included in a modeling exercise in the Netherlands to derive a Cd SQS that is dependent on soil pH because the latter parameter sensitively affects the BAF. Ingwersen and Streck (2005) demonstrated that for a number of field crops grown on sandy soil the regression relationship between soil and tissue Cd was specific to year, and they speculated that the reason was year-to-year variability in water use by the crops, as Cd is readily translocated from roots to shoots in the plant's transpiration stream. Whatever the approach (soil extract or empirical models), it is imperative that there is sufficient field validation across soil types, seasons, etc. The uncertainty of the BAF can be taken forward by using the lower confidence limits of the regression model predictions.

Atmospheric contribution is a second factor explaining the weak association between crop and soil contamination. Even washed crops may contain TEs that were deposited from air on the plants during plant growth, and it is possible that dissolved metals are taken up through pores on the leaf surface. There are, however, few studies on the contribution of airborne TEs to tissue concentration because it requires either experimental studies with filtered air control treatments or isotopic studies. A combination of these techniques showed that the fraction of metal in wheat (flour) derived from atmosphere was 21% for Cd and more than 90% for Pb for wheat grown in a rural area (Dalenberg and Vandriel 1990). It is unclear how large these fractions are for oxyanions such as As, Se, and Mo. The food chain contamination through atmosphere is a factor that complicates the risk assessment because the premise of deriving an SQS is that the human exposure relates to the soil contamination. Logically, the contribution is large if soil contamination is low compared to air contamination and conversely. Without doubt, it can be stated that the large contribution of atmospheric contribution to crop Pb concentrations is a factor by which food chain exposure to Pb is difficult, if not impossible, to predict from soil Pb contamination (cf. Zhao et al. 2004).

3.5.4 Soil-Crop-Animal and Soil-Animal transfer

Animal products (meat and dairy products) are obviously another source of dietary exposure to the TEs. Before starting to identify the transfer data, it is recommended that a sensitivity analysis is performed with background data from food surveys on the importance of these products in the dietary intake. For several metals, for example, it is known that animal products usually contribute little to the total intake unless offal is consumed. Lack of suitable data is the most challenging factor for this assessment. As with soil-crop transfer, bioavailability factors come into play, related to the speciation of the TE in the food or ingested soil and with large uncertainties on actual intake rate of the food items (including soil ingestion rate) of the animals. Some models break down this analysis by supplying different TFs (crop-animal, animal-meat, meat-milk). Without going into detail, we recommend that background data of the TEs in soil and in food items must be used to validate the predictions of these models with observed concentrations in the animal products.

3.6 STATISTICAL CHARACTERIZATION OF EXPOSURE

Quantitative evaluation of uncertainty can be undertaken using either deterministic or probabilistic approaches. Deterministic approaches using point estimates are widely practiced by the Joint World Health Organization/Food and Agriculture Organization (WHO/FAO) Expert Committee on Food Additives (WHO/FAO 2006) and are considered to be suitable at the international level to assess exposure to various contaminants such as pesticide and veterinary drug residues in connection with the establishment of Codex maximum residue limits of these contaminants in foods. The widespread use of deterministic approaches for estimating exposure rather than probabilistic estimates of exposure does, however, limit the characterization of exposure from various sources. The applicability of probabilistic approaches to the assessment of contaminants, including TEs, in food has been considered previously by the Joint Expert WHO/FAO Committee on Food Additives (WHO/FAO 2006), who concluded that data availability is often inadequate to support the use of probabilistic assessments, thus this approach is not, generally, realistically feasible at the present time. Uncertainty factors (UFs) (also called safety factors and assessment factors) are used in the development of guideline values for many media (food, water, and soil) to account for uncertainties in the database, including extrapolations of toxicity from animal studies and variability within humans, which result in some uncertainty about risk. The application of UFs is entrenched in toxicological risk assessment and regulatory practices worldwide but is not necessarily applied consistently, as a well-documented and disciplined basis for the selection of these factors is not always apparent.

3.6.1 Probabilistic versus Deterministic Assessments

Where adequate data are available, probabilistic exposure assessment approaches offer a departure from a single value estimate of each exposure parameter as required for a deterministic exposure assessment. However, clear distinctions must be made in a probabilistic exposure assessment between variability and uncertainty. It is generally not appropriate to mix variability and uncertainty in a parameter because this results in a standard deviation that is inflated beyond that expected and inappropriately affects the predicted tails of the exposure distribution. Possible approaches for probabilistic exposure assessment include the following:

1) Perform a probabilistic assessment only on those parameters that are variable, using best estimates of the parameter distributions. For example, daily average soil ingestion varies among children, as exposure frequency may also vary. This will provide a distribution of likely exposures across a population as variability of at least some exposure parameters is reasonably well understood, but it will not address uncertainty.
2) Perform a probabilistic assessment based only on those parameters that are uncertain. For example, the average concentration of a contaminant in an exposure area is uncertain, where the extent of uncertainty is in part a function of the amount of sampling. This will provide a distribution of potential exposures to a selected individual in the population, such as the average or maximally exposed individual. Little is known about the bounds of uncertainty for many exposure parameters.
3) Perform a 2-phase probabilistic assessment where both variability and uncertainty are assessed but in a "nested" fashion. For example, first, a simulation is done where the variability in all exposure parameters in considered, and then a second simulation is done where uncertainty in all exposure parameters is considered. The outcome of this type of analysis is a distribution of (uncertain) risks for each member of a variable population. That is, the simulation can yield the average and 95th percentile estimates of risk to the median population, and the average and 95th percentile estimates of risk to the 95th percentile of the population.

There are (at least) 2 difficulties in moving from a deterministic to a probabilistic analysis. The first is that some measure of transparency is lost owing to the complexity of the calculations, and the second is that the results can be difficult to explain to stakeholders. A deterministic calculation can generally be easily checked by an outside party, while complex probabilistic simulations require expertise to program, cannot easily be checked, and, because the results are probabilistic, do not always yield the same answer each time a simulation is run, at least at the tails of the predicted distribution. Both the complexity and the difficulty in checking the calculations can make the communication of the results to stakeholders difficult.

It can also be difficult to choose the point within the distributions of variability and uncertainty that will be used for risk management decisions such as setting an SQS. For example, do we choose to base SQSs such that the 95th percentile of the population has no more than a 50% chance of exceeding a risk threshold or the 50th percentile of the population has no more than a 95% chance of exceeding the same risk threshold, or some other combination? The difficulty in making this choice should not alone be a barrier to use of probabilistic exposure assessment because in fact we are making the same choice in the use of a deterministic risk assessment; the choice is just less transparent.

In the deterministic approach, various assumptions are included in the analysis. For example, all food contains the TE at the maximum level or only food based on a specific land use pattern contains residues of the TE. When the dietary exposure to a given TE exceeds the TDI, a more refined estimation could be conducted, applying either deterministic or, where supported by adequate data, probabilistic methods. Refinements that may be considered could include 1) the proportion of food that may contain the TE, 2) residue and monitoring data to confirm the presence of the TE in the food, and 3) market basket survey data to better understand exposure potential and the potential impact of processing and food preparation. The impact of these various alternative assumptions on the outcome can then be determined. This results in a range of possible outcomes, which can then be used in the assessment of the overall uncertainty.

In general, an iterative evaluation of uncertainty is undertaken, in which progressively more of the uncertainties are quantified, until there is sufficient confidence in the overall outcome of the assessment.

3.6.2 Uncertainty Factors

The derivation and application of UFs have been substantially influenced by historical practice (Ritter et al. 2007). Additional factors are sometimes included to account for other areas of uncertainty. Application of UFs may include extrapolating subchronic data to anticipated chronic exposure, use of a lowest observable adverse effect level (LOAEL) instead of a no observable adverse effect level (NOAEL) and other database deficiencies such as high mortality in laboratory chronic feeding studies. In some cases, the magnitude of the UF may be exaggerated by consideration of several database deficiencies individually rather than by lumping these together as a single database deficiency factor. The default value attributed to each UF is generally a factor of 3 or 10; however, little comprehensive guidance is available to inform to the selection of UFs.

As an example, the UFs for the derivation of drinking water standards in a number of areas of uncertainty are as follows (Beck and Dourson 1993):

- All uncertainties have been resolved, generally 1;
- Any 1 area, generally 10;
- Any 2 areas, generally 100;

- Any 3 areas, generally 1000;
- Any 4 areas, generally 3000;
- Any 5 areas, generally 10,000.

However, the USEPA indicates that the convention of multiplying individual UFs is modified when there are 4 or more areas of uncertainty to avoid overlap and overprotection because of a UF potentially reaching 100,000 (Dourson et al. 1996). If there are 4 or more areas of uncertainty, then 2 areas of uncertainty will be combined within one 10-fold factor (Dourson et al. 1996). For example, if there are 4 full areas of uncertainty, instead of obtaining a total UF of 10,000 (as could be the case by 10^4), a total UF of 3000 is used owing to overlap in UFs. In 2002, the USEPA recommended limiting the total UF applied for any particular chemical to no more than 3000 and avoiding the derivation of a reference value that involves application of the full 10-fold UF in 4 or more areas of extrapolation (USEPA 2002).

A decision tree has been developed by WHO to provide guidance for selection of appropriate UFs to account for the range of uncertainty encountered in the risk assessment process. Recent development of a series of decision trees by WHO to derive chemical-specific adjustment factors for inter- and intraspecies variability may present an opportunity for a more systematic approach for the identification of evidence-based UFs. The WHO (2005) decision tree approach is given below (Figure 3.1).

Identification and evaluation of the assumptions and sources of uncertainty in a risk assessment are important in ensuring transparency and promoting consistency in the risk assessment. Appendix III of the Proposed Draft Codex Working Principles for Risk Analysis states that: "Any constraints, uncertainties and assumptions and their impact on the risk assessment should be documented in a transparent manner, including constraints that are likely to influence the quality of the risk assessment. Expression of uncertainty or variability in risk estimates may be qualitative or quantitative.

It is clearly not feasible, or indeed necessary, to quantify all sources of variability and uncertainty in a risk assessment. If qualitative consideration of a source of uncertainty provides sufficient confidence (e.g., the assessment is clearly conservative) to enable risk managers to reach a decision, then quantitative assessment would be unnecessary.

3.7 ESSENTIAL TEs

Essential TEs require special considerations in the establishment and interpretation of SQSs. By definition, the absence or deficiency of such substances in the human diet produces functional and/or structural abnormalities in processes as diverse as reproduction, immune system function, nervous system function, and growth (WHO 1996). The functions fulfilled by an essential metal cannot be replaced by any other substance, although symptoms of deficiency can generally be reversed upon a resumption of nutritionally adequate levels, depending on the length and severity of the deficiency.

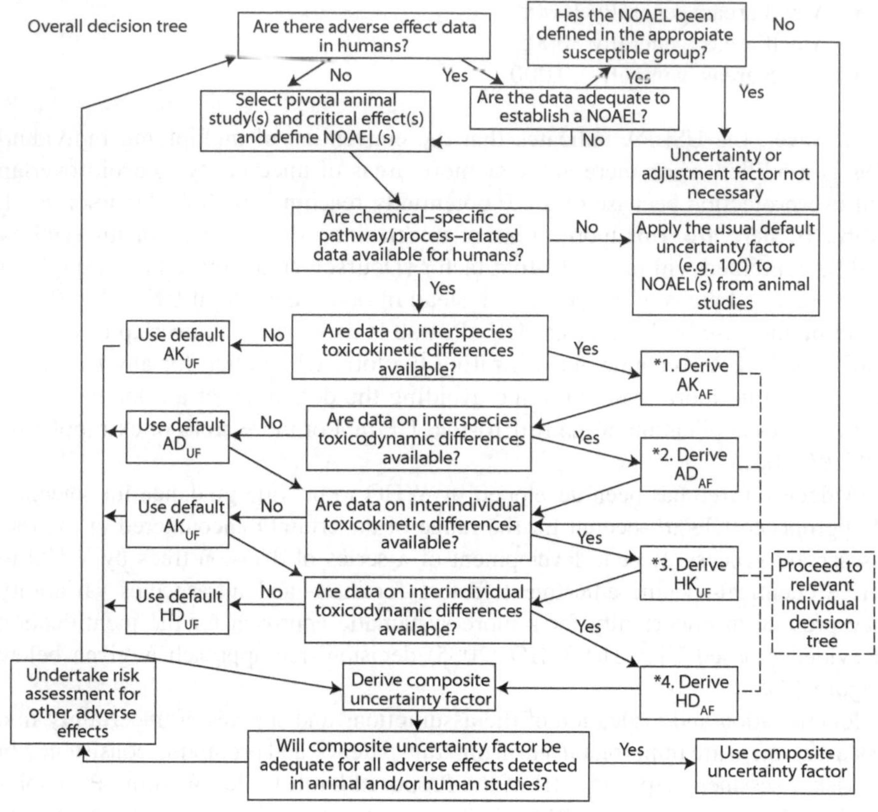

FIGURE 3.1 WHO and IPCS overall decision tree.

As established by WHO (2002), TEs considered as essential for human health include iron, zinc, copper, chromium, cobalt, molybdenum, selenium, and iodine. For most of these metals, specific functions (e.g., as structural components of enzymes) have been identified. Definitive data for essentiality are less robust for a second group of elements classified as "probably essential for humans" and include silicon, manganese, nickel, boron, and vanadium.

3.7.1 Homeostasis and the Setting of SQSs

In higher animals and plants, essential TEs are under the control of homeostatic mechanisms that include regulation of absorption rates, levels of tissue retention, and rates of excretion (WHO 2002). Homeostatic mechanisms enable adaptation to varying nutrient intakes to ensure a safe and optimum systemic supply of nutrients for the performance of essential metabolic functions. For example, under conditions of poor nutrition, the uptake of zinc from the gastrointestinal tract is elevated and can approach or exceed 40% uptake efficiency (Hunt et al. 2008). Under conditions of higher exposure, downmodulation of transport occurs and uptake declines to approximately 9% (Babcock et al. 1982). The efficiency of homeostatic processes and nutritional requirements vary within human populations owing to both genetic and lifestyle factors. The derivation of recommended daily intakes thus entails a process of estimating variability in intake requirements and recommending intake levels that satisfy the nutritional requirements of most individuals in a population. The required uptake levels for an essential metal are normally established from observational studies in humans and estimates then made of the RDI that will satisfy uptake needs (Institute of Medicine 2001). RDI development must thus give due consideration to the bioavailability of metals in different foods and the dietary patterns of subsets of the general population. For example, diets rich in meat protein contain significant levels of metal in a bioavailable form. In contrast, strict vegetarians may consume grains rich in essential minerals, but the minerals are bound to substances (e.g., phytates) that restrict bioavailability and thus inhibit uptake. The intake requirements for a vegetarian are accordingly higher than those of an individual consuming a meat protein-rich diet. RDIs must thus be viewed within the context of the dietary habits of the population to which they will be applied.

Excess exposure to essential TEs (even the essential elements) can produce symptoms of toxicity (e.g., impaired reproductive function) but normally manifests itself when levels of exposure exceed that which can be accommodated by homeostatic mechanisms (WHO 2002). In a number of instances, toxicity from excess exposure to essential elements is in fact a reflection of perturbations in the metabolism of other TEs that may share common aspects of homeostatic control. For example, 1 of the earliest adverse effects associated with excess exposure to zinc is impaired copper uptake and the induction of mild copper deficiency (WHO 2001).

3.7.2 Dose-Response Relationships for Essential Elements

The dose response for health effects associated with essential TEs is generally best considered as a U-shaped dose response curve (WHO 2002). Low levels of exposure are associated with health impairments related to deficiency. Higher levels of exposure provide intake optimal for health with an acceptable range of oral intake that reflects the range of exposures within the capacity of homeostatic control mechanisms. The width of this range of acceptable intake varies with the element under consideration. For some elements the difference between intake levels that produce deficiency and those that produce toxicity can be quite broad. For others, the acceptable range of oral intake can be quite narrow or nonexistent. The latter situation is generally observed when genetic heterogeneity with human populations produces individuals with different uptake requirements for optimal health. In some cases these differences can be extreme as exemplified by Wilson's disease, a genetic disorder characterized by impaired hepatic copper transport and/or excretion that results in excessive copper accumulation in the liver (Brewer 2000). In the absence of therapeutic intervention, increasing levels of hepatic copper will ultimately result in significant impairment of liver function. Dietary copper levels required for adequate nutrition in most individuals can thus be lethal for the Wilson's disease patient. Finally, toxicity occurs at oral intakes greater than those that would be considered sufficient.

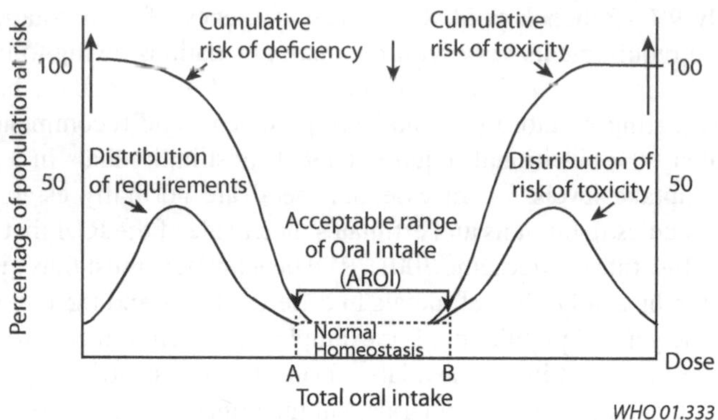

FIGURE 3.2 A model U-shaped dose response.

A model U-shaped dose response is depicted below (Figure 3.2) (derived from WHO 2002) and illustrates the complexity inherent in defining population-based requirements for essential element intake as well as defining upper exposure limits such as TDIs.

The application of traditional toxicological UFs to the derivation of TDIs for essential TEs is often problematic. For example, optimal zinc intake for an adult human consuming a mixed diet is generally 12 to 15 mg d^{-1} (Food and Drug Administration 2007). The acceptable range of oral intake appears to extend to approximately 60 mg d^{-1} at which point biochemical alterations are observed (e.g., inhibition of erythrocyte superoxide dismutase [e-SOD]) that may be indicative of perturbations in copper metabolism (Agency for Toxic Substances and Disease Registry [ATSDR] 2005). Different agencies have regarded this change as either a LOAEL (USEPA 2003) or a NOAEL (ECB, 2004) that provides the basis for the establishment of a TDI. However, the application of traditional safety factors (e.g., a factor of 10) to this upper exposure limit is problematic in that the resulting TDI (6 mg d^{-1}) would induce deficiency and have deleterious impacts upon health in the majority of the population. Alternate safety factors are thus required but vary as a function of the expert judgment of the Agency establishing the TDI. For example, USEPA opted to apply a safety factor of 3 to derive a RfD of 21 mg d^{-1} for an adult male. While this RfD is higher than the RDI, the nutritional intake requirements of some sectors of the population (e.g., vegetarians) may be higher (Hunt et al. 2008); thus for these groups, the RfD and RDI might be similar. The recent European Union Zn risk assessment interpreted superoxide dismutase (SOD) inhibition as a NOAEL, applied a safety factor of 1, thereby judging 60 mg d^{-1} to be the upper range of acceptable exposure (Netherlands 2004). The WHO (2004) also recognized the distinctly different dietary patterns that may affect zinc bioavailability but recommended that the average level of zinc intake for adults in a population should not exceed 45 mg d^{-1}.

3.7.3 Interaction of Essential and Nonessential TEs

The essentiality of some elements merits consideration in evaluating potential impacts of nonessential toxic elements. In a number of instances, the carrier systems involved in homeostatic control of essential TEs uptake provide transport mechanisms for the uptake of nonessential elements. Conditions of zinc (Chaney et al. 2001) or iron (Vahter et al. 2007) deficiency thus seem to facilitate the uptake of cadmium into the body. Calcium (Manton et al. 2003), and to a certain extent, iron deficiency (Wasserman et al. 1994) can facilitate the uptake of lead. Adequate nutrition for essential TEs can reduce the uptake of nonessential TEs by both homeostatic down-modulation of active transport processes and/or by competing with nonessential elements for binding to uptake transport proteins within the gastrointestinal tract. More precision can be obtained in assessing risks associated with nonessential elements when such nutritional interactions are recognized and factored into the identification of exposure scenarios that may increase or decrease uptake (and thus risk). For example, risk factors for Itai-Itai disease (a cadmium-related disease affecting both renal function and bone metabolism in women) included both dietary deficiencies for essential elements and increased risk for multiparous women due to presumed increased nutritional demands for calcium during pregnancy (Vahter et al. 2007).

Recognition of the role of transport mechanisms for essential elements in the uptake of nonessential elements also assists in defining the nonlinear toxicokinetics that can characterize uptake in humans. For example, initial uptake of a nonessential element can occur through both efficient active transport and less efficient passive diffusion processes. Active transport processes are, by their nature, saturable and capacity limited. Uptake of nonessential elements can, as a function of increasing dose, exhibit nonlinearity (e.g., decreased uptake efficiency) as uptake through passive diffusion becomes the primary mode of element uptake (WHO 1996). It further follows that essentiality has an impact on SQSs that might be developed for the protection of crops or elements of ecosystems. As with human health, consideration should be given to the restrictions imposed by essentiality in the application of safety factors.

3.8 BIOAVAILABILITY AND BIOACCESSIBILITY

Some TEs in ingested soil can comprise a risk to animals or humans, particularly at elevated concentrations, and in bioavailable forms; it is especially well studied for Pb and As but may also be important for Cd, Ni, F, Hg, and possibly other TEs. Soil ingestion circumvents the soil-plant barrier whereby limited plant uptake limits significant exposure (As, Ni, Pb, and perhaps Cr). Incidental soil ingestion by children is an important pathway in assessing public health risks associated with exposure to TE-contaminated soils (Chaney and Ryan 1994; Dudka and Miller 1999). Exposure must be quantified considering the magnitude, frequency, and duration of exposure for the receptors and pathways selected for quantitative evaluation. For incidental ingestion, the following formula can be used to quantify average daily chemical intake (USEPA 1989):

$$\mathrm{CDI} = \frac{(\mathrm{CS})(\mathrm{IR})(\mathrm{CF})(\mathrm{FI})(\mathrm{EF})(\mathrm{ED})}{(\mathrm{BW})(\mathrm{AT})}, \tag{3.1}$$

where

CDI = Chemical daily intake (mg kg d^{-1})
CS = Chemical concentration in soil (mg kg^{-1})

IR = Ingestion rate (mg soil d^{-1})
CF = Conversion factor (10^{-6} kg mg^{-1})
FI = Fraction ingestion from contaminated source (unitless)
EF = exposure frequency (days y^{-1})
ED = Exposure duration (y)
BW = Body weight (kg)
AT = Average time (period over which exposure is averaged, days)

The CS variable is the total TE content in soil. In quantifying metal intake by the above formula, the underlying assumption is that all of the TE measured by the total metal analysis is absorbed. However, there is an inherent problem with this assumption. For an adverse health effect to be realized, the chemical toxicant (in this case, the TE) must be dissolved for absorption to occur. Several forms of TEs in soil may not be soluble under conditions similar to those of the human gut. The combination of various chemical species with different soil and/or solid matrices of TE produces a wide range of TE solubility. Most metal and metalloid sulfides, for example, are less soluble than their respective oxidized compounds: for example, the solubility of As_2S_3 in water is 0.005 g L^{-1}, while the solubility of As_2O_3 is 37 g L^{-1}. These differences have a significant impact on the dose absorbed from ingestion of contaminated soil. The bioavailable TE is defined as "the fraction of an ingested dose that crosses the gastrointestinal epithelium and becomes available for distribution to internal target tissues and organs" (USEPA 2007a, p. 2). Bioavailability of a TE in soil can be divided into 2 kinetic steps: dissolution of TE in gastrointestinal fluids and absorption across the gastrointestinal epithelium into the blood stream. The variability of geochemical forms of TEs in soil combined with dissolution chemistry and biological absorption processes in the gastrointestinal tract results in a complex system. Controlled dosing studies are required to accurately determine the bioavailability of TE in this complex system.

The enteric bioavailability of a TE can be described in absolute (ABA) or relative (RBA) terms. ABA (also referred to as the oral absorption fraction AF_o) is equal to absorbed dose/ingested dose:

$$ABA = \frac{\text{Absorbed dose}}{\text{Ingested dose}}. \tag{3.2}$$

Relative bioavailability is the ratio of the ABA of the TE present in some test material (i.e., soil) compared to the ABA of the TE in the reference material used to determine TDI in Integrated Risk Information System (IRIS, USEPA), typically the chemical dissolved in water or some fully soluble form that completely dissolves in the digestive tract:

$$RBA = \frac{\text{ABA (test material)}}{\text{ABA (reference material)}}. \tag{3.3}$$

As an example, if 100 μg of a TE dissolved in drinking water were ingested and a total of 80 μg entered the body, the ABA would be 0.80 (80%). Likewise, if 100 μg of a TE contained in soil were ingested and 20 μg entered the body, the ABA for the soil would be 0.20 (20%). If the TE dissolved in water was used as the reference substance, the RBA would be 0.20/0.80 = 0.25 (25%), and if the reference substance and its delivery to the animal were the same as in the IRIS database, then the toxicity prediction by IRIS could be adjusted.

This simple model assumes toxicokinetics of the ingested TE that may not in actual fact be so simple. For example, in most animals and humans, absorbed As is excreted mainly in the urine. Consequently, the urinary excretion fraction (UEF), defined as the amount excreted in the urine divided by the amount ingested, is a reasonable approximation of the oral absorption fraction or ABA. However, this ratio will underestimate total absorption because some absorbed As is excreted back into the intestine via the biliary mechanism, and some absorbed As enters tissue compartments (e.g., liver, kidney, skin, and hair) from which it is cleared very slowly. Therefore the UEF should not be equated with the absolute absorption fraction. Absolute bioavailability (ABA = AF_o) of As from a test material can be estimated from the ratio of UEFs of As from test material compared to intravenously dosed As using the same species as the study that formed the basis of the IRIS guideline. Then RBA of 2 orally dosed materials (e.g., a test soil and sodium arsenate) can be calculated from the ratio of their UEFs, a calculation which is independent of the extent of tissue binding and biliary excretion, and where D is the dose of As and K_u is the fraction of absorbed As (or any TE) excreted in the urine:

$$\text{RBA (test versus ref)} = \frac{AF_o(\text{test})}{AF_o(\text{ref})} = \frac{(D)(AF_o\ \text{test})(K_u)}{(D)(AF_o\ \text{ref})(K_u)} = \frac{\text{UEF (test)}}{\text{UEF (ref)}}. \qquad (3.4)$$

The mass of As excreted in urine by each animal over each collection period (48 hours) is plotted as a function of the amount administered (μg As 48 hours^{-1}), and the slope of the best fit straight line through the data is used as the best estimate of the UEF. The relative bioavailability of As in test material is calculated as

$$\text{RBA} = \frac{\text{UEF (test material)}}{\text{UEF (reference material)}}, \qquad (3.5)$$

where sodium arsenate is used as the soluble reference form of As.

As noted in Equation (3.5), each RBA value is calculated as the ratio of 2 slopes (UEFs), each of which is estimated by linear regression through a set of data points, and each of which has associated uncertainty, described by the standard error of the mean for the slope parameter.

Further discussion is available on the derivation and use of soil Pb RBA (Casteel et al. 1996, 2006) and soil As RBA (Basta et al. 2001a). Human subjects are not commonly used in dosing studies for ethical reasons; more typically, appropriate animal models are used to determine bioavailability of TEs in contaminated soils. Recently reviewed in vivo models used to measure bioavailable TEs (i.e., Pb, As) include juvenile swine, monkey, rabbit, and dog (Weis and LaVelle 1991, Casteel et al. 1996, 2006; Valberg et al. 1997; Ruby et al. 1999; Roberts et al. 2002, 2007). In these in vivo dosing trials, soil TE bioavailability is evaluated by measuring TE in urine, blood, feces, and/or storage tissues (bone, skin, nails, and hair). Juvenile swine, and less frequently monkey, have been used for determining site-specific bioavailability of soil TEs for use in risk assessment at Superfund sites because they are remarkably similar to humans with respect to their digestive tract, nutritional requirements, bone development, and mineral metabolism (Dodds 1982). Young swine are considered to be good physiological models for gastrointestinal absorption in children (Weis and LaVelle 1991; Casteel et al. 1996). Other benefits of juvenile swine include the economics of husbandry, ease of dose delivery, and the concern of animal rights' groups regarding animal model selection. Furthermore, the swine model for bioavailability determinations has gained acceptance as a method to determine TE RBA (USEPA 2007b). Dosing trials using primates and swine are expensive, however. Recently, USEPA scientists have developed a faster, less expensive in vivo adult mouse model for evaluating the bioavailability of As-contaminated soils. The bioavailability of As in a standard reference soil determined by the mouse model agreed with results determined by the significantly more expensive swine model (Bradham et al. 2008). Inbred strains of laboratory mice are well characterized physiologically and can be easily manipulated experimentally (e.g., altered dietary components, altered genotype). A large body of data is available on the absorption, metabolism, disposition, and excretion of inorganic As in the mouse that is germane to evaluating the similarities between mice and humans. The USEPA has developed a Physiologically Based Pharmacokinetic (PBPK) model for As based on mice (El-Masri and Kenyon 2008; Evans et al. 2008).

3.8.1 Use of In Vitro Gastrointestinal Methods to Estimate TE Bioavailability

To avoid the difficulty and expense associated with in vivo trials, research has been directed toward the development of in vitro methods that simulate human gastrointestinal conditions for the estimation of bioaccessible TEs in soils. Several of these methods have been reviewed (Rodriguez et al. 1999; Ruby et al. 1999; Oomen et al. 2003). Regardless of the in vitro gastrointestinal (IVG) method used to measure bioaccessible TE in soil, the measurements must be well correlated with in vivo estimates of bioavailable TE. According to USEPA Guidance (USEPA 2007a, p. 13):

> In the case that a validated *in vitro* method is used to estimate bioavailability, it is recommended that the protocol specified in the methodology be followed for making the extrapolation from *in vitro* data to *in vivo* values. That is, there is no *a priori* assumption that all validated *in vitro* methods must yield results that are identical to *in vivo* values. Rather, it is assumed that a mathematical equation will exist such that the *in vitro* result (entered as input) will yield an estimate of the *in vivo* value (as output).

At a minimum, the in vitro method must be correlated with TE RBA measured by an acceptable animal model. However, in vitro methods that are able to predict bioavailable TE with an estimate of uncertainty are highly desirable. Because of the cost of animal dosing trials, few studies comparing IVG methods with animal models have been conducted. Ruby et al. (1996) developed of an IVG method called the physiologically based extraction test (PBET) to determine bioaccessible Pb. PBET uses synthetic gastric solution containing pepsin, and the pH is adjusted using HCl to 1.5 or greater (depending on the conditions being simulated); the soil is incubated with the gastric solution at 37°C under argon, maintaining pH with HCl or bicarbonate. PBET-extracted TE was strongly correlated with bioavailable Pb in soil measured by immature swine: Schroder et al. (2004) reported bioaccessible Pb measured by a similar IVG method, the Ohio State University (OSU) IVG, was correlated with Pb RBA using immature swine ($r = 0.89$, $p < 0.01$. Drexler and Brattin (2007) developed their relative bioavailability leaching procedure (RBALP), which uses 0.4 M glycine to buffer the solution at pH 1.5 to mimic fasting stomach pH; the extraction is conducted at 37 °C for 1 hour, using 1 g dry soil and/or dust per 100 mL extraction fluid. Bioaccessible Pb measured by RBALP was strongly correlated ($r = 0.91$) with Pb RBA determined using the immature swine model. The RBALP has been adopted as USEPA Guidance for adjusting oral bioavailability of Pb in soil in human health risk assessments (USEPA 2007b). Several IVG methods for measuring bioaccessible As have been described (Basta et al. 2007; Lowney et al. 2007). Basta et al. (2001b) and Rodriguez et al. (1999) reported a strong correlation ($r = 0.91$, $p < 0.01$) of bioaccessible As measured by the OSU IVG method, with As RBA determined by immature swine, for 14 contaminated soils. Juhasz et al. (2007) reported a strong correlation ($r = 0.96$, $p < 0.01$) of bioaccessible As measured by PBET, with RBA As determine using swine, for 12 contaminated soils. Lowney et al. (2007) reported a correlation between bioaccessible As and RBA As in Cynomolgus monkey. Ruby et al. (1996) compared bioaccessible As measured using PBET with RBA As using rabbit and Cynomolgus monkey, for 3 contaminated soils. Bioaccessible As (i.e., PBET) overpredicted RBA As. The small number of soils prevented a thorough comparison of As PBET with As RBA.

Most studies show that As bioavailability in contaminated soil can be much lower (typically less than 50%) than the bioavailability of the inorganic As (i.e., soluble sodium arsenate) used for assessing risk from As in drinking water (Ruby et al. 1999). Bioavailability of As in contaminated soil relative to sodium arsenate (i.e., relative bioavailability, RBA) ranged from 0 to 98% with a median value of 35.5% for 16 contaminated soils and media fed to rabbits (Ruby et al. 1999) and from 4.07 to 42.9% with a median value of 25.5% for 14 contaminated soils and media fed to swine (Basta et al. 2001b).

Rodriguez et al. (2003) and Yang et al. (2002) reported that arsenic in contaminated soil that was associated with amorphous (i.e., reactive) Fe oxide minerals was not bioavailable. Beak et al. (2006) found Fe oxides surfaces in ferrihydrite greatly reduced As bioaccessibility to less than 5% RBA. The speciation of As, determined using extended X-ray absorption fine structure X-ray absorption near-edge spectroscopy, was determined to be strong binuclear bidentate bonding with the Fe oxide surface.

Co-ingestion of food and contaminated soil has been shown to decrease bioavailability of soil Pb to humans as compared to soil ingestion without food (Maddaloni et al. 1998). Therefore fasted conditions are considered conservative estimates of Pb RBA and should be used for measuring bioaccessible Pb by IVG methods. While fasted conditions have been adopted by IVG methods used to measure bioaccessible As, this might not result in a conservative estimate of As RBA because phosphate associated with diets may increase As bioaccessibility and perhaps RBA (Basta et al. 2007).

Studies on the application of IVG methods to TEs other than Pb and As are limited. Bioaccessible Cd, measured by the OSU IVG method, was correlated with Cd RBA ($r = 0.91$) derived from Cd in the renal cortex of immature swine (Schroder et al. 2003). However, data for Cd and other TE studies are very limited; research is underway to explore the relationships between bioaccessible Ni and Ni RBA.

3.9 EFFECTS OF CHARACTERIZATION

For comparison to the estimate of exposure, there must be a mathematically described relationship to a human health endpoint. While human health risk assessment is often considered to be simple because only 1 species is considered, parameterization of the effects endpoint in many cases is coarse and thus may result in quite imprecise estimate of risk.

3.9.1 Benchmark Dose versus NOAEL/LOAEL

The benchmark dose (BMD) is a dose of a chemical or agent that is predicted to cause a given effect or response based on the experimentally derived dose response relationship. The response corresponding to the BMD is the benchmark response (BMR); there can be more than 1 BMD for a TE, each of which would correspond to a different endpoint. The BMD is derived by modeling the dose response relationship in the range of the experimental data, then, based on the model, extrapolating or interpolating to a dose corresponding to the expected environmental exposure. For many dose response relationships, the extrapolation is to the lower doses more typical of environmental exposure (USEPA 1995, 2000). The lower confidence limit (generally the 95th percentile), known as the benchmark dose limit (BMDL) is used as the BMD. The BMD/BMDL is not necessarily an observed dose response; rather, it is an interpolation or extrapolation of the relationship between the experimentally defined doses and the observed responses.

The BMD approach to characterizing toxicity can be used for any chemical or agent and for any adverse effect. It can be used as an alternative to single experimentally defined doses such as the NOAEL or LOAEL as the basis for toxicity values that inform regulation. The BMD/BMDL is preferred to using NOAELs or LOAELs (providing that there are sufficient data to construct a statistically and biologically significant dose response relationship (more than 5 or 6 experimental doses) that includes variation in response) as unlike NOAELs/LOAELs, it is independent of the doses chosen for experimentation, and its precision is estimable. For effects where no threshold is believed to exist (e.g., most carcinogens) the BMDL may also serve as a point of departure (POD) for linear low-dose extrapolation (i.e., where the low-dose extrapolation begin, it is most often the upper bound on an observed incidence or on an estimated incidence from a dose-response model, http://www.epa.gov/IRIS/gloss8_arch.htm). After the POD has been derived, a linear extrapolation is made from high dose levels in the range of experimental doses to very low environmentally relevant doses. Acceptable dose levels are generally in the range of those associated with excess cancer risks of 1 in 1×10^{-6} or 1 in 1×10^{-5}. Examples of nonthreshold toxicity values that are used to relate excess cancer risk with dose include cancer slope factors (CSFs). The BMD is the maximum likelihood estimate rather than a lower confidence limit (BMDL). The BMR can be expressed as extra or additional risk above background. Generally, using the extra risk approach is more conservative. Using the BMDL as the POD provides for using more of the information contained in the dose response relationship as the basis for beginning the linear extrapolation. The BMDL is more precisely defined than using a predefined experimental dose. Where data are adequate, using the BMDL as the POD is preferred to using an experimental dose, where data are adequate.

3.9.2 Bridging Ambient Exposure to Literature Doses

PBPK modeling is a method that simulates biological processes (e.g., absorption, distribution, metabolism, elimination). It can be used to predict the relationship between the dose administered in a toxicological study (e.g., chemical concentration in drinking water or in breathing zone air (i.e., proximity to the nostrils) and the resulting internal dose in tissues or organ. It is particularly useful in human health risk assessment, as it can be used to predict target organ doses (dose where damage occurs) in study animals and predicts an equivalent target organ dose in humans then relates the dose to a human administered dose. This provides a more precise understanding of whole animal bioassay results and better information on which to evaluate potential human exposure and risk.

Many regulatory bodies prefer to use more chemical-specific information and science for these extrapolations (e.g., USEPA 1994a, 1994b, 2002, 2005a, 2005b) when data are appropriate. If a PBPK model has not been developed or is not used, then the administered dose is adjusted using default approaches, parameters, and assumptions that are not chemical specific. The following are examples of commonly used default scaling approaches:

1) A scaling factor of body weight (USEPA 2005a) is often used to convert an animal dose to a human equivalent dose for carcinogens. Scaling factors are generally not used for noncarcinogenic effects. Species' differences are more commonly addressed by applying an additional UF of 10 to the NOAEL, LOAEL, or BMDL to derive the toxicity value. A method was proposed (USEPA 2002) in which a UF of 3 would be applied along with a dosimetric adjustment factor (DAF) that is based on body weight parameters. The DAF would be equivalent to 0.25(animal body weight)/0.25(human body weight).

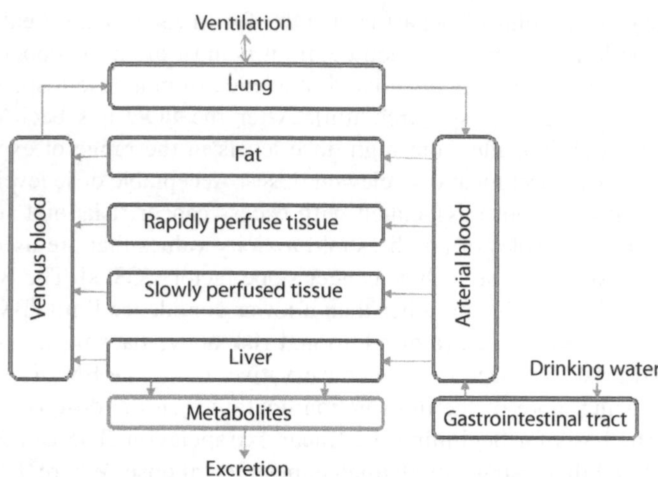

FIGURE 3.3 Generic representation of a PBPK model.

2) For inhalation exposure, converting from an animal dose to a human dose is based on the chemical form (e.g., particle, gas), whether it causes effects in the lungs or systemically, and the region of the lung that is exposed to the chemical. The general equation for calculation of the human equivalent dose (HED) is $NOAEL_{[HED]}$ (mg/m^3) = $NOAEL_{[ADJ]}$ (mg/m^3) x DAF (USEPA 1994a). The $NOAEL_{[ADJ]}$ is the dose from the study adjusted to a weekly average. The DAF is based on chemical-physical properties of the chemical, the region of the lung exposed to the chemical, and whether the chemical exerts its effect at the route of entry or systemically. This approach is used for both noncancer and cancer endpoints.

The PBPK model reflects the chemical's characteristics and describes how it behaves in the body. Some models are simple with few compartments; some are much more complex. What follows is a generalized depiction of a PBPK model (Figure 3.3).

Dose estimation and scaling based on a PBPK model more precisely define the dose response relationship than if based on the administered dose. Clearly, using as much chemical-specific, test animal-specific, and human-specific information that is available is more rigorous and more scientifically defensible than using a generic scaling factor or UFs. This is recognized in agency guidance (e.g., USEPA 2005a); however, agencies are sometimes reluctant to use the available PBPK models. This may be because models are simulating very complex processes and may be complex or there is a reluctance to change from what has historically been done. There is always a level of uncertainty in these models as there is in all aspects of the risk assessment process. Clearly, models must be critiqued, dissected, and evaluated and uncertainties recognized and accounted for before being used in human health risk assessment. However, these uncertainties need to be clearly defined and weighed against the uncertainties associated with using generic scaling factors and UFs.

3.9.3 Sensitive Subpopulations

Human populations are diverse and include groups that may be more sensitive to the effects of environmental contaminants than the general population. These sensitive subpopulations are accounted for in the development toxicity values and when evaluating potential risks and hazards. However, sometimes default approaches do not adequately address these populations. Potentially sensitive populations may include the very young (e.g., fetuses, infants, children), pregnant women, the elderly, and chronically and acutely ill people. In addition, groups may be more sensitive to the effects of a particular chemical because of physiology, genetics, activities, and/or cultural practices. It is important that population characteristics as well as mode of action and target organ be considered when evaluating potential risks and hazards. A common approach to accounting for the difference between the general population and sensitive population for threshold effects (generally noncancer end points) is to reduce the toxicity value by an additional UF of 10. However, this approach should be modified if it does not appropriately address the hazard. Children may be more sensitive to some harmful effects and less sensitive to others than adults. Children's bodies are changing, growing, and maturing. They are building bones, muscles, tissues, and organs. Their metabolism, breathing rate, and hormonal state are very different from adults. When children are identified as particularly susceptible to the effects of a chemical or agent, it needs to be addressed. Ideally, it should be accounted for in the toxicity values used to evaluate risk. The following provides examples of where special consideration to the sensitivity of children or other subpopulations has been identified and approaches adjusted to account for this sensitivity.

3.9.3.1 Children Subpopulations

Very young children (younger than 6 years of age) have been identified as the most sensitive population to the effects of inorganic lead, in the form of damage to the CNS, which can result in diminished mental ability and capacity for an entire lifetime. No toxicity threshold has been identified, so, even though this is a noncancer effect, exposure to lead is most commonly regulated assuming there is no threshold. The Integrated Exposure Uptake Biokinetic model is often used to predict blood lead concentrations in young children as a result of exposure. Young children are also are at increased risk to the effects of lead because the body may transport lead free ion (Pb^{2+}) as it transports calcium free ion (Ca^{2+}), something that is enhanced by poor nutrition and/or low calcium intake. Lead can accumulate in the growth plates at the end of long bone in highly exposed children. It may remain for years then remobilize because of poor nutrition (e.g., low calcium intake), disease, or even when there is a high need for calcium such as during pregnancy. The extent of this release is a function of many variables, including the internal concentrations, age, and general health.

Exposure during early life stages to mutagenic carcinogens may pose a greater risk than comparable exposure during later life stages. Cancer is a collection of diseases characterized by cellular growth that does not respond to normal homeostatic controls. For regulatory purposes, carcinogens have been divided by some agencies into those where there may be a risk associated with any level of exposure (no threshold) (such as those caused by mutations) and those where harmful effects are not expected if the dose is sufficiently low (a threshold) (such cell damage and increased cellular proliferation). If data are available, life stage specific CSFs should be developed. If data are lacking, 1 approach proposed for accounting for this potential susceptibility is applying an age-dependent adjustment factor to CSFs (CSF–toxicity value). The higher the CSF, the more potent the carcinogen, the lower the dose associated with the level of acceptable risk. USEPA has recommended the following in their Supplemental Cancer Guidelines (USEPA 2005b):

- From birth to younger than 2 years: 10-fold adjustment factor (CSF × 10).
- Between 2 years and younger than 16 years: 3-fold adjustment factor (CSF × 3).
- From 16 years to adulthood: No adjustment factor.

3.9.3.2 Adult Subpopulations

As people age, they are more likely to suffer from chronic conditions that can make them more vulnerable to the effects of particular chemicals or agent. Conditions like emphysema, chronic obstructive pulmonary disease, and asthma can cause increased sensitivity to respirable particulate. Any disease of a particular organ or tissue would increase sensitivity to a chemical that causes damage to that site. If a population engages in a behavior that leads to a higher than expected background exposure to a particular TE, acceptable levels of exposure may need to be adjusted. If the population is the entire regulated community, then this may be more appropriately addressed in the toxicity value. Otherwise, this needs to be addressed during the risk assessment. For example, some North American Indian tribes in the United States and Canada still rely on the local environment for much of their food and water, and thus diets may include a very high intake of local fish and game. If these "country foods" have elevated concentrations of TEs such as Hg and Cd, then background exposures will be higher than assumed for the general population. This should be accounted for during the risk assessment by modification of the estimates for background exposure and adjustment of the TDI to reflect this higher background exposure.

3.10 SUMMARY AND CONCLUSIONS

This chapter has explored the reasons behind variations in the derivation methodologies used for establishing SQS between several standards-setting jurisdictions. It is clear that exposure estimates present some of the greatest challenges to scientists, practitioners, and regulators in this field; this provided the greatest focus for this chapter.

Although brief, the review of how background exposure is considered and calculated in the setting of SQS illustrates whether or not it is subtracted from the TDI, and how it is parameterized, varies among jurisdictions. Some of this variance is policy rather than science; for example, the decision as to how much of the TDI can be attributed to contaminated soil has little to no scientific foundation. While harmonizing the method for consideration of background exposure would improve the rationale behind SQSs, it is not clear from this analysis that the lack of harmony among methods makes a particularly large contribution to the overall variance in SQSs.

None of the jurisdictions included in this survey estimate inhalation exposure relative to particle size; rather, it is simply based on total particulate mass. It seems clear that the potential for UFP to be the significant source of exposure, proportionally greater than its mass would suggest, the particle size distribution should be considered in human exposure assessment.

With a very limited number of elements, nickel being the prime example, there is a robust body of evidence on inhalation exposure and its consequences. Most of this information has been derived from industrial exposure, where there is likely to be substantially greater exposure than for the general population and at the upper boundary of the nonlinear absorption curve. In this portion the rate of change of retention with respect to exposure is very much less than occurs at the lower portion of the curve. It is also notable that the endpoints of inhalation toxicity are often noncarcinogenic.

A flexible, tiered approach to the evaluation of uncertainty is therefore recommended: 1) all identifiable sources of uncertainty should be reported; 2) all of these should be evaluated qualitatively; and 3) where uncertainty remains as to their qualitative impact on the assessment, these should be quantified to the extent necessary to provide sufficient reassurance to enable risk management decisions.

The sources of uncertainty need to be identified in both the toxicology and the exposure assessment. An illustration of the application of such an approach can be found in the work of the WHO/Food and Agriculture Organization (2006). All steps in the assessment should be considered systematically and evaluated qualitatively. One approach to this is to estimate the likely magnitude of uncertainty and direction of the impact of each source of uncertainty on the potential health outcome, using a subjective scoring system of pluses and minuses, with a range as necessary, i.e., –, – –, – – –, +, ++, +++, and combinations of these. Similarly, the uncertainty in the overall evaluation is assigned a score or range of scores based on subjective evaluation of the individual scores. It is often useful to include a tabular summary of the qualitative uncertainty analysis.

Conventional risk assessment strategies and UFs can be applied to nonessential elements in the derivation of TDIs that, in turn, may serve as the basis for SQSs that limit human exposure that may occur through either uptake into food crops or direct ingestion of soils and dusts. However, when applied to essential TEs, traditional UFs and risk assessment procedures may lead to the derivation of TDIs that would adversely affect health through the induction of nutritional deficiencies. Traditional toxicological evaluations for essential elements must thus be expanded to include both the high exposure levels that may produce toxicity and a formal determination of the intake levels required for optimal health so as to ensure that excessive UFs are not applied for TDI derivation. Nutritional status for essential elements can also modify (increase or decrease) the uptake of nonessential toxic elements. Recognition of TE nutrition as a modulating factor can assist in the identification of sensitive subpopulations and the derivation of strategies to mitigate increased risk. They further provide a mechanistic foundation that can assist in the identification of likely interactions between substances and thus guide the formulation of strategies for the study of the impacts of mixtures.

Finally, adjustments to TE exposure for bioavailability refine the prediction of adverse health effects based on toxicity studies using soluble forms of TEs. However, adjustments for bioavailability are relevant only if the relative bioavailability in soil or soil + food are significantly different that IRIS default values. For example, the default reference value for Pb is an RBA of 0.6 or 60% (USEPA 2004). Adjustments should be considered for Pb exposure if the soil or soil + food RBA differs markedly from 60%. Soil + food has been shown to decrease Pb RBA to less than 3% to humans (Maddaloni et al. 1998). Insoluble forms of soil Pb have reported RBA values for Pb of less than 10%. These RBA Pb values are significantly lower than the default Pb and would warrant RBA adjustments to Pb exposure and subsequent adjustments in SQS for Pb. If validated bioaccessibility and/or bioavailability TE measurements are used for SQS calculation, then the UF relating to data relevance should be removed or reduced from 10 to a smaller value. Fasted conditions should be used to determine TE bioavailability and/or bioaccessibility in the nondietary exposure. Because food has been shown to greatly decrease Pb bioaccessibility (Schroder et al. 2003) and Pb bioavailability (Maddaloni et al. 1998) and phosphate in food may increase As bioaccessibility (Basta et al. 2007), fed conditions (i.e., soil and food) should be used to determine TE bioavailability in the dietary pathway. The many determinations of bioavailability that are now becoming available for TEs in various soils and animal models raises the possibility that deterministic values could be replaced in the near future by probabilistic values for use in risk assessment. This would be consistent with the earlier discussion in this chapter, suggesting that deterministic values for soil ingestion be replaced by such metrics as the 95th percentile of the literature median or mean values.

REFERENCES

Adams ML, Zhao FJ, McGrath SP, Nicholson FA, Chambers BJ. 2004. Predicting cadmium concentrations in wheat and barley grain using soil properties. J Environ Qual 33:532–541.

Agency for Toxic Substances and Disease Registry (ATSDR). 2005. Toxicological profile for zinc. Atlanta (GA): US Department of Health and Human Services.

Asgharian B, Menache MG, Miller FJ. 2004. Modeling age-related particle deposition in humans. J Aerosol Med 17(3):213–224.

Babcock AK, Henkin RI, Aamodt RL, Foster DM, Berman M. 1982. Effects of oral zinc loading on zinc metabolism in humans: II. In vivo kinetics. Metabolism 31:335–347.

Basta NT, Foster JN, Dayton EA, Rodriguez RR, Casteel SW. 2007. The effect of dosing vehicle on arsenic bioaccessibility in smelter-contaminated soils. J Environ Health Sci A42:1275–1281.

Basta NT, Ryan L, Chaney RL. 2005. TE chemistry in residual-treated soil: key concepts and metal bioavailability. J Environ Qual 34(1):49–63.

Basta NT, Rodriguez RR, Casteel SW. 2001a. Bioavailability and risk of arsenic exposure by the soil ingestion pathway. In: Frankenberger WT, editor, Environmental chemistry of arsenic. New York: Marcel Dekker. 117–139.

Basta NT, Rodriguez RR, Casteel SW. 2001b. Development of chemical methods to assess the bioavailability of arsenic in contaminated media. Final Report. Washington (DC): National Center for Environmental Research, USEPA. 1–40.

Beak DG, Basta NT, Scheckel KG, Traina SG. 2006. Bioaccessibility of arsenic (V) bound to ferrihydrite using a simulated gastrointestinal system. Environ Sci Technol 40:1364–1370.

Beck B, Dourson M. 1993. How toxicity data are used in the process of hazard identification and dose-response assessment. Paper presented at Society of Toxicology 32nd Annual Meeting, March 14–18, 1993, New Orleans (LA).

Brewer G. 2000. Minireview: recognition, diagnosis, and management of Wilson's disease. Proc Soc Exp Biol Med 223:39–46.

Bradham K, Thomas D, Basta N, Van de Wiele T, Hughes M, Scheckel K, Harper S, Creed J. 2008. Determine bioavailability and bioaccessibility of arsenic in soil and develop arsenic speciation methods. Draft FY06 ORD Pilot Report. Paper presented at 2005 USEPA Technical Review Meeting, November 14–18, Safety Harbor (FL).

Brown SL, Chaney RL, Angle JS, Ryan JA. 1998. The phytoavailability of cadmium to lettuce in long-term biosolids amended soils. J Environ Qual 27:1071–1078.

Canadian Council of Ministers of the Environment (CCME). 2006. A protocol for the derivation of environmental and human health soil quality guidelines. http://www.ccme.ca/assets/pdf/sg_protocol_1332_e.pdf (accessed 19 June 2010).

Calabrese EJ, 2003. Letter from Edward Calabrese to K. Holtzclaw re: Soil Ingestion Rates. July 23. It can be found in Appendix E of the following document: Comments of the General electric company on the U.S Environmental Protection Agency's human health risk assessment for the Housatonic River Site- rest of river, prepared for General Eletric by AMEC Earth and Environmental, Inc. and BBL Sciences, July, 28, 2003. http://www.epa.gov/ne/ge/thesite/restofriver/reports/final_hhra/comments/generalelectric/AttachE.pdf

Calabrese EJ, Pastides H, Barnes R, Edwards C, Kostecki PT. 1989. How much soil do young children ingest: an epidemiologic study. In: Petroleum contaminated soils. Chelsea (MI): Lewis Publishers. p 363–397.

Casteel SW, Cowart RP, Henningsen GM, Hoffman E, Brattin WJ, Starost F, Payne JT, Stockham SL, Becker SV, Turk JR. 1996. A swine model for determining the bioavailability of lead from contaminated media. In: Tumbleson ME, Schook LD, editors, Advances in swine in biomedical research. Vol. 2, Proceedings of International Symposium on Swine in Biomedical Research, 637–646. New York (NY): Plenum P. p 637–646.

Casteel SW, Weis CP, Henningsen GW, Brattin WJ. 2006. Estimation of relative bioavailability of lead in soil and soil-like materials using young swine. Environ Health Perspect 114:1162–1171.

Chalupa DC, Morrow PE, Oberdörster G, Utell MJ, Frampton MW. 2004. Ultrafine particle deposition in subjects with asthma. Environ Health Perspect 112(8):879–882.

Chaney RL. 1980. Health risks associated with toxic metals in municipal sludge. In: Bitton G, Damron BL, Edds GT, Davidson JM, editors, Sludge — health risks of land application. Ann Arbor (MI): Ann Arbor Science. 59–83.

Chaney RL, Ryan JA. 1994. Risk based standards for arsenic, lead and cadmium in urban soils. Frankfurt (Germany): DECHEMA. 130 p.

Contaminated Land Exposure Assessment (CLEA). 2008. Human health toxicological assessment of contaminants in soil. Science Report SC050021/SR2. Bristol (UK): Environment Agency.

Daigle CC, Chalupa DC, Gibb FR, Morrow PE, Oberdörster G, Utell MJ, Frampton MW. 2003. Ultrafine particle deposition in humans during rest and exercise. Inhal Toxicol 15(6):539–552.

Dalenberg JW, Vandriel W. 1990. Contribution of atmospheric deposition to heavy-metal concentrations in field crops. Neth J Agric Sci 38:369–379.

Davis S, Waller P, Buschbon R, Ballou J, White P. 1990. Quantitative estimates of soil ingestion in normal children between the ages of 2 and 7 years: population based estimates using aluminum, silicon, and titanium as soil tracer elements. Arch Environ Health 45:112–122.

Dodds JW. 1982. The pig model for biomedical research. Fed Proc 41:247–256.

Dourson ML, Felter SP, Robinson D. 1996. Evolution of science-based uncertainty factors in noncancer risk assessment. Regul Toxicol Pharmacol 24:108–120.

Drexler JW, Brattin WJ. 2007. An in vitro procedure for estimation of lead relative bioavailability: with validation. Human Ecol Risk Assess 13:383–401.

Dudka S, Miller WP. 1999. Permissible concentrations of arsenic and lead in soils based on risk assessment. Water Air Soil Pollut 113:127–132.

Efroymson RA, Semple BE, Suter GW III. 2001. Uptake of inorganic chemicals from soil by plant leaves: regressions of field data. Environ Toxicol Chem 20:2561–2571.

Elert M, Jones C, Norman F. 1997. Development of generic guideline values: model and data used for generic guideline values for contaminated soils in Sweden. Stockholm (Sweden): Swedish Environmental Protection Agency, Naturvardsverket.

El-Masri H, Kenyon E. 2008. Development of a human physiologically based pharmacokinetic (PBPK) model for inorganic arsenic and its mono- and di-methylated metabolites. J Pharmacokinet Pharmacodyn 35:31–68.

Elleman CJ, Barclay WS. 2004. The M1 matrix protein controls the filamentous phenotype of influenza A virus. Virology 321(1):144–153.

Eriksson J, Öborn I, Jansson G, Andersson A. 1996. Factors influencing cadmium content in crops — Result from Swedish field investigations. Swed J Agric Res 26:125–133.

European Chemicals Bureau, 2004. Zinc Distearate. Part II – Human Health. CAS No: 557-05-1 & 91051-01-3 EINECS No: 209-151-9 & 293-049-4. Summary Risk Assessment Report. Special Publication I.04.118.

Evans M, Dowd S, Kenyon E, Hughes M, El-Masri H. 2008. A physiologically based pharmacokinetic model for intravenous and ingested dimethylarsinic acid in mice. Toxicol Sci 104(2):250–260.

Food and Drug Administration. 2007. Food labeling: revision of reference values and mandatory nutrients. Fed Reg 101(212):62,149–175.

Geiser M, Rothen-Rutishauser B, Kapp N, Schürch S, Kreyling W, Schulz H, Semmler M, Hof VI, Heyder J, Gehr P. 2005. Ultrafine particles cross cellular membranes by nonphagocytic mechanisms in lungs and in cultured cells. Environ Health Perspect 113(11):1555–1560.

Ghio AJ, Cohen MD. 2005. Disruption of iron homeostasis as a mechanism of biologic effect of ambient air pollution particles. Inhal Toxicol 17(13):709–716.

Holmes NS, Morawska L, Mengersen K, Jayaratne ER. 2005. Spatial distribution of submicrometre particles and CO in an urban microscale environment. Atmos Environ 39(22):3977–3988.

Hunt JR, Beiseigel JM, Johnson LK. 2008. Adaptation in human zinc absorption as influenced by dietary zinc and bioavailability. Am J Clin Nutr 87:1336–1345.

Hyde DM, Blozis SA, Avdalovic MV, Putney LF, Detorre R, Quesenberry NJ, Singh P, Tyler NK. 2007. Alveoli increase in number but not size from birth to adulthood in rhesus monkeys. Am J Physiol Lung Cell Mol Physiol 293(3):L570–L579.

Ingwersen J, Streck T. 2005. A regional-scale study on the crop uptake of cadmium from sandy soils: measurement and modeling. J Environ Qual 34:1026–1035.

Institute of Medicine. 2001. Dietary reference intakes for vitamin A, vitamin K, arsenic, boron, chromium, copper, iodine, iron, manganese, molybdenum, nickel, silicon, vanadium and zinc. Washington (DC): National Academy Pr.

International Commission on Radiological Protection (ICRP). 1994. Human respiratory tract model for radiological protection. Oxford (UK): Pergamon Pr.

Juhasz AL, Smith E, Weber J, Rees M, Rofe A, Kuchel T, Sansom L, Naidu R. 2007. Comparison of *in vivo* and *in vitro* methodologies for the assessment of arsenic bioavailability in contaminated soils. Chemosphere 69:961–966.

Kim CS, Jaques PA. 2004. Analysis of total respiratory deposition of inhaled ultrafine particles in adult subjects as various breathing patterns. Aerosol Sci Technol 38(6):525–240.

Kim CS, Jaques PA. 2005. Total lung deposition of ultrafine particles in elderly subjects during controlled breathing. Inhal Toxicol 17(7–8):387–399.

Li N, Sioutas C, Cho A, Schmitz D, Misra C, Sempf J, Wang M, Oberley T, Froines J, Nel A. 2003a. Ultrafine particulate pollutants induce oxidative stress and mitochondrial damage. Environ Health Perspect 111(4):455–460.

Li F, Okazaki M, Zhou Q. 2003b. Evaluation of Cd uptake by plants estimated from total Cd, pH and organic matter. Bull Environ Contam Toxicol 71:714–721.

Lowney YW, Roberts S, Saikat S. 2007. Arsenic bioaccessibility testing using various extraction methods: results and relation to relative oral bioavailability as measured in the Cynomolgus monkey. Abstract 461. International Society for Exposure Assessment, October 14–18, Durham (NC).

Maddaloni M, LoIacono N, Manton W, Blum C, Drexler J, Graziano J. 1998. Bioavailability of soilborne lead in adults, by stable isotope dilution. Environ Health Perspect 106(Suppl. 6):1589–1594.

Manton WI, Angle CR, Stanek KL, Kuntzelman D, Reese YR, Kuehnemann TJ. 2003. Release of lead from bone in pregnancy and lactation. Environ Res 92:139–151.

Massaro D, DeCarlo Massaro G. 2007. Developmental alveologenesis: longer, differential regulation and perhaps more danger. Am J Physiol Lung Cell Mol Physiol 293(3):L568–L569.

McLaughlin MJ, Zarcinas BA, Stevens DP, Cook N. 2000. Soil testing for heavy metals. Commun Soil Sci Plant Anal 31(11–14):1661–1700.

McLaughlin MJ, Whatmuff M, Warne M, Heemsbergen D, Barry G, Bell M, Nash D, Pritchard D. 2006. A field investigation of solubility and food chain accumulation of biosolid-cadmium across diverse soil types. Environ Chem 3:428–432.

Morawska L, Ristovski Z, Moore MR. 2004. Health effects of exposure to ultrafine particulates. Report to the Australian Department of the Environment and Heritage, ISBN 0-642-55055-7, 1-207.

Nan Z, Zhao C, Li J, Chen F, Sun W. 2002. Relations between soil properties and selected heavy metal concentrations in spring wheat (*Triticum aestivum* L.) grown in contaminated soils. Water Air Soil Pollut 133:205–213.

Nel A, Xia T, Mädler L, Li N. 2006. Toxic potential of materials at the nanolevel. Science 311(5761):622–627.

Nemmar A, Hoet PHM, Vanquickenborne B, Dinsdale D, Thomeer M, Hoylaerts MF, Vanbilloen H, Mortelmans L, Nemery B. 2002. Passage of inhaled particles into the blood circulation in humans. Circulation 105(4):411–414.

Netherlands. 2004. 2004 European Union risk assessment report on zinc metal, zinc(ii)chloride, zinc sulphate, zinc distearate, zinc oxide, trizinc bis(orthophosphate). Report prepared by Netherlands, RIVM on behalf of the European Union.

Oberdörster G, Oberdörster E, Oberdörster J. 2005. Nanotoxicology: an emerging discipline evolving from studies of ultra-fine particles. Environ Health Perspect 113(7):823–839.

Olsson I-M, Eriksson J, Öborn I, Skerfving S, Oskarsson A. 2005. Cadmium in food production systems: a health risk for sensitive population groups. Ambio 34(4–5):344–351.

Oomen AG, Tolls J, Sips AJAM, Van den Hoop MAGT. 2003. Lead speciation in artificial human digestive fluid. Arch Environ Contam Toxicol 44:107–115.

Provoost J, Cornelis C, Swartjes F. 2006. Comparison of soil clean-up standards for trace elements between countries: why do they differ? J Soils Sediments 6(3):176–181.

Ritter L, Totman C, Krishnan K, Carrier R, Vézina A, Morisset V. 2007. Deriving uncertainty factors for threshold chemical contaminants in drinking water. J Toxicol Environ Health Part B 10:527–557.

Roberts SM, Munson JW, Lowney YW, Ruby MV. 2007. Relative oral bioavailability of arsenic from contaminated soils measured in the Cynomolgus monkey. Toxicol Sci 91(1):281–288.

Roberts SM, Weimar WR, Vinson JRT, Munson JW, Bergeron RJ. 2002. Measurement of arsenic bioavailability in soil using a primate model. Toxicol Sci 67:303–310.

Rodriguez RR, Basta NT, Casteel SW, Pace LW. 1999. An in vitro gastrointestinal method to estimate bioavailable arsenic in contaminated soils and soil media. Environ Sci Technol 33:642–649.

Rodriguez RR, Basta NT, Casteel SW, Armstrong FP, Ward DC. 2003. Chemical extraction methods to assess bioavailable arsenic in soil and solid media. J Environ Qual 32(3):876–884.

Roy M, Courtay C. 1991. Daily activities and breathing parameters for use in respiratory tract dosimetry. Radiat Prot Dosimetry 35(3):179–186.

Ruby MV, Davis A, Schoof R, Eberle S, Sellstone CM. 1996. Estimation of lead and arsenic bioavailability using a physiologically based extraction test. Environ Sci Technol 30:422–430.

Ruby MV, Schoof R, Brattin W, Goldade M, Post G, Harnois M, Mosby DE, Casteel SW, Berti W, Carpenter M, Edwards D, Cragin D, Chappell W. 1999. Advances in evaluating oral bioavailability of inorganics in soil for use in human health risk assessment. Environ Sci Technol 33:3697–3705.

Sapoval B, Filoche M, Weibel ER. 2002. Smaller is better-but not too small: a physical scale for the design of the mammalian pulmonary acinus. Proc Natl Acad Sci USA 99(16):10,411–416.

Scarpelli EM. 2003. Physiology of the alveolar surface network. Comp Biochem Physiol Part A 135(1):39–104.

Schroder JL, Basta NT, Casteel SW, Si J. 2003. An in vitro method to estimate bioavailable cadmium in contaminated soil. Environ Sci Technol 37:1365–1370.

Schroder JL, Basta NT, Casteel SW, Evans TJ, Payton ME, Si J. 2004. Validation of the in vitro method to estimate bioavailable lead in contaminated soils. J Environ Qual 33:513–521.

Stanek EJ, Calabrese EJ. 1995. Daily estimates of soil ingestion in children. Environ Health Perspect 103(3):276–285.

Stanek EJ, Calabrese EJ. 2000. Daily soil ingestion estimates for children at a Superfund site. Risk Anal 20(5):627–635.

Strum M, Cook R, Thurman J, Ensley D, Pope A, Palma T, Mason R, Michaels H, Shedd S. 2006. Projection of hazardous air pollutant emissions to future years. Sci Total Environ 336(2–3):590–601.

Sznitman J, Heimsch F, Heimsch T, Rusch D, Rosgen T. 2007. Three-dimensional convective alveolar flow induced by rhythmic breathing motion of the pulmonary acinus. J Biomech Eng 129(5):658–665.

USEPA. 1989. Risk assessment guidance for superfund (RAGS). Vol. I, Human health evaluation manual (Part A). Report EPA/540/1-89/002. Washington (DC): Office of Emergency and Remedial Response.

USEPA. 1994a. Methods for derivation of inhalation reference concentrations and application of inhalation dosimetry. Report EPA/600/8-90/066F. Washington (DC): USEPA.

USEPA. 1994b. Interim policy for particle size and limit concentration issues in inhalation toxicity: notice of availability. Fed Reg 59:53799.

USEPA. 1995. Use of the benchmark dose approach in health risk assessment. Report EPA/630/R-94/007. Washington (DC): USEPA, National Center for Environmental Assessment.

USEPA. 2000. Benchmark dose technical guidance document. Report EPA/630/R-00/001. http://www.epa.gov/ncea/iris/backgrd.htm (accessed 19 June 2010).

USEPA. 2002. A review of the reference dose and reference concentration processes. Report EPA-630/P-02/00F. Washington (DC): USEPA, Office of Research and Development, Office of Health and Environmental Assessment. http://cfpub.epa.gov/ncea/cfm/recordisplay.cfm?deid=55365.

USEPA. 2003. Zinc and zinc compounds. http://www.epa.gov/ncea/iris/subst/0426.htm (accessed 19 June 2010).

USEPA. 2004. United States user's guide/technical background document for deriving Preliminary Remediation Goals (PRGs). http://www.epa.gov/Region9/waste/sfund/prg/files/02userguide.pdf (accessed 19 June 2010).

USEPA. 2005a. Guidelines for carcinogenic risk assessment. Report EPA/630/P-03/001F. Washington (DC): USEPA.

USEPA. 2005b. Supplemental guidance for assessing susceptibility from early-life exposure to carcinogens. Report EPA/630/R-03/003F. Washington (DC): USEPA.

USEPA. 2007a. Guidance for evaluating the oral bioavailability of metals in soils for use in human health risk assessment. Report OSWER 9285.7-80. Washington (DC): USEPA.

USEPA. 2007b. Estimation of relative bioavailability of lead in soil and soil-like materials using in vivo and in vitro methods. Report OSWER 9285.7-77. Washington (DC).: USEPA.

USEPA. 2008. Integrated science assessment for particulate matter. Report ISA: EPA/600/R-08/139 Annexes: EPA/600/R-08/139A. Research Triangle Park (NC): USEPA. 938 p.

Vahter M, Akesson A, Liden C, Ceccatelli S, Berglund M. 2007. Gender differences in the disposition and toxicity of metals. Environ Res 104:85–95.

Valberg PA, Beck BD, Bowers TS, Keating JL, Bergstrom PD, Boardman PD. 1997. Issues in setting health-based cleanup levels for arsenic in soil. Reg Toxicol Pharmacol 26:219–229.

Van Holderbeke M, Cornelis C, Bierkens J, Torfs R. 2007. Review of the soil ingestion pathway in human exposure assessment. Final Report. 194 p. VITO, Mol, Belgium. Study in support of the BeNeKempen project-subproject on harmonization of the human health risk assessment methodology. http://www.vito.be/HomepageAdmin/Home/WetenschappelijkOnderzoek/MilieurisicoEnGezondheid/

Wasserman GA, Graziano JH, Factor-Litvak P, Popovac D, Morina N, Musabegovic A, Vrenezi N, Capuni-Paracka S, Lekic V, Preteni-Redjepi E, Hadzialjevic S, Slavkovich V, Shrout JP, Stein Z. 1994. Consequences of lead exposure and iron supplementation on childhood development at age 4 years. Neurotoxicol Teratol 16:233–240.

Weis CP, LaVelle JM. 1991. Characteristics to consider when choosing an animal model for the study of lead bioavailability. Chem Spec Bioavail 3:113–119.

WHO. 1996. Trace elements in human nutrition and health. Geneva (Switzerland): World Health Organization.

WHO. 2001. Zinc. International Programme on Chemical Safety Environmental Health Criteria 221. Geneva (Switzerland): World Health Organization.

WHO. 2002. Principles and methods for the assessment of risk from essential trace elements. International Programme on Chemical Safety Environmental Health Criteria 228. Geneva (Switzerland): World Health Organization.

WHO. 2004. Vitamin and mineral requirements in human nutrition. Geneva (Switzerland): World Health Organization.

WHO. 2005. Chemical-specific adjustment factors (CSAFs) for interspecies differences and human variability in dose/concentration-response assessment: guidance document for use of data in dose/concentration response assessment. International Programme on Chemical Safety, Harmonization Project Document 2. Geneva (Switzerland): World Health Organization.

WHO/Food and Agriculture Organization. 2006. Updating the principles and methods of risk assessment. Rome (Italy): World Health Organization/Food and Agriculture Organization.

Yang JK, Barnett MO, Jardine PM, Basta NT, Casteel SW. 2002. Adsorption, sequestration, and bioaccessibility of As (V) in soils. Environ Sci Technol 36:4562–4269.

Zeman KL, Bennett WD. 2006. Growth of the small airways and alveoli from childhood to the adult lung measured by aerosol-derived airway morphometry. J Appl Physiol 100(3):965–971.

Zhao FJ, Adams ML, Dumont C, McGrath SP, Chaudri AM, Nicholson FA, Chambers BJ. 2004. Factors affecting the concentrations of lead in British wheat and barley grain. Environ Pollut 131:4611–4668.

4 Implementation and Use of Terrestrial Standards for Trace Elements

Graham Merrington, Ilse Schoeters, Michael Warne, Beverley Hale, Victor Dries, Co Molenaar, Jaana Sorvari, Jussi Reinikainen, Seung-Woo Jeong, Chris Oates, Gladys Stephenson, Lucia Buvé, John Chapman, Diane Heemsbergen, Randy Wentsel, Andreas Bieber, and Wang Guoqing

4.1 INTRODUCTION

In using and implementing soil quality standards (SQSs) there is a requirement to consider more than the concentration of a chemical in soil (Royal Commission on Environmental Pollution 1998). Generally, an SQS will be a component of a broader decision making process or framework and will be a chemical-specific value ascribed to a particular ecological target or receptor (e.g., humans, soil fauna), protection level, or designated soil or land use. SQSs are one tool within a regulatory framework and, unlike aquatic quality standards, are rarely used as mandatory pass or fail criteria. Nevertheless, comparing the numerical values of SQSs from across jurisdictions, without reference to the context in which they are to be used or how they have been derived, is a meaningless exercise.

The objectives of this chapter are to review some of the key aspects of the use and implementation of SQSs for trace elements (TEs). Specifically, we propose defensible but readily useable and practical approaches to the assessment of environmental and human health risks from TEs in soils. This has been facilitated through the sharing of regulatory experiences and recent developments on the implementation and use of SQS in order to identify technically robust and broadly applicable methods.

The environmental regulation of TEs presents several fundamental challenges that are not generally encountered in considering organic chemicals. These challenges have recently been summarized by the USEPA (2007a) and include the following:

- All TEs are naturally occurring constituents in the environment and vary in concentration across geographic regions.

- All environmental media have naturally occurring mixtures of TEs, and they are often introduced into the environment as mixtures.
- Some TEs are essential for maintaining the proper health of humans, animals, plants, and microorganisms. For such TEs, adverse effects on humans and biota can occur through environmental concentrations being too low or too high. This contrasts with nonessential TEs and all synthetic organic compounds that can cause harmful effects by being present in environmental media in excess.
- TEs, unlike many organic chemicals, are neither created nor destroyed by biological or chemical processes, although these processes can transform metals or metalloids from one species to another (valence states) and can convert them between inorganic and organic forms.
- The absorption, distribution, transformation, and excretion of TEs within an organism depend on the element involved, the form of the element or elemental compound, and the organism's ability to regulate or store the element.

This chapter is divided into 9 parts. Section 4.2 briefly reviews the uses and purpose of SQSs before Section 4.3 considers the frameworks in which SQSs for TEs are used by regulatory agencies in Europe, North America, and Australasia. Sections 4.4 and 4.5 attempt to identify "best practice" in terms of the means of implementing SQSs for TEs that account for ambient background concentrations (ABCs), availability, and mixtures, with examples of how such practice can be validated through the use of field data. In Sections 4.6 to 4.9 some of the practical issues related to the derivation and implementation of SQSs for TEs are discussed including accounting for mixtures, monitoring, data acquisition, and communication to stakeholders. Finally, in Section 4.10 we provide conclusions and several practical recommendations that would lead to the implementation of scientifically robust methods to identify potential risks from TEs in soils.

4.2 THE USE OF SOIL QUALITY STANDARDS

"Soil quality standard" is a term whose definition and meaning will depend on the legal framework and jurisdiction in which it is being used. An extensive overview of definitions and meanings in the different European Union (EU) countries is given by Carlon (2007) and these are largely true for other jurisdictions also. The term "quality standard" for the terrestrial compartment has previously been defined (Crane et al. 2009) as

- a numerical value related to soil,
- a threshold for decisions, and
- related to a target (such as ecological receptors or humans) or a designated use, and having a specific protection or trigger level.

SQSs can be classified as being of 3 main types: negligible risk values, trigger values, and action values (Merrington et al. 2009):

1) Negligible risk values indicate the concentration at or below which it is deemed that the risk to human health or ecosystems is negligible. These can be used to define the long-term environmental objective.
2) Trigger values indicate the concentration that if exceeded is deemed to pose potentially unacceptable risks to human health, ecosystems, or groundwater. Exceeding the trigger value initiates further investigations and site-specific risk assessment (e.g., National Environment Protection Council [NEPC] 1999a).
3) Action values (also sometimes termed intervention values) indicate the concentration which if exceeded initiates some action such as remediation, or restrictions on the use of the land/soil.

These types of SQSs differ in the level of risk deemed acceptable, reflecting the specific purpose of the SQSs. Generally, the derivation of negligible risk values aims at excluding any type of adverse effect even in the most sensitive land use. It is characterized by a high level of conservatism. Alternatively, the derivation of action values aims at preventing significant adverse effects occurring and is characterized by a lower level of conservatism (Carlon 2007).

Additional types of SQSs, which have different purposes to the above, include

1) Screening values which are used to identify contaminants of potential concern or priority action;
2) Criteria for the reuse of excavated soil, waste material, manufactured soil in soil, and land management (e.g., British Standards Institution [BSI] 2007);
3) Values that indicate the state of the soil resource at a regional or national level and aid in planning decisions which are not related to contamination (Arshad and Martin 2002; Sparling et al. 2003).

Along with the diversity in definitions and meanings described above, Carlon (2007) also noted diversity in the receptors being protected by different SQSs. Most countries that have SQSs aim to protect human health. Some countries also have SQSs that aim to protect the terrestrial ecosystem, i.e., soil and terrestrial organisms. Much less frequently, SQSs aim to protect groundwater or surface water (Carlon 2007).

Generally, quality standards for contaminants are derived separately for different compartments of the environment (e.g., surface water, groundwater, soil, and sediment). This approach is largely due to a compartmentalized approach to dealing with environmental risks. However, it has been established that some TEs may move from one compartment to another and can be transported globally (e.g., mercury). The problem with deriving environmental quality standards (EQS) on a compartmental basis is that meeting the EQS in one compartment may lead to exceedance of the EQS in another compartment. For example, the EQS for Cu in water could lead to the need for greater removal rates in water treatment plants, which could lead to an exceedance of the EQS in sewage sludge and possibly soils to which the biosolids are applied. To overcome this compartmental approach and the inherent problems of this approach, the Dutch developed the "Simple Box" multicompartment transport model (Brandes et al. 1996), which is based on earlier work on fugacity modeling by Mackay (2001). However, this method of harmonizing EQS across different compartments has not subsequently been adopted by the Dutch to derive SQSs. A recent example where potential off-site impacts of contaminants were considered in deriving EQSs is an Australian project to derive quality standards for contaminants in fertilizers (Sorvari et al. 2008, 2009) that has been submitted to the Primary Industries Standing Committee and the Environment and Protection Heritage Council of Australia. These quality standards for fertilizers (fertilizer contaminant trigger values) were derived by calculating back from EQSs for soil, groundwater, surface water, sediment, fish, seafood, and livestock to the concentration in fertilizers that would not cause an exceedance of any of the above EQSs after 100 years of continual application of fertilizers.

An SQS can cover both ecological and human health risk through the use of a single value, such as the Canadian Environmental Quality Guidelines (Canadian Council for Ministers for the Environment [CCME] 2006) and the Dutch intervention and target values (van Vlaardingen et al. 2005). In such cases, although values for each receptor are derived, the lower of these 2 values is generally the final SQS that is implemented (Ministry for the Environment 2003; CCME 2006). Generally, most jurisdictions have separate SQSs to protect human health or the environment. However, even if 2 values are derived and applied separately, the lowest SQS will be the one that drives the risk assessment for a particular site at the initial screening tier.

Much of the discussion in the following sections relates to contaminated land, but it should be stressed that the scope of implementation and use of TE SQSs for soils and soil-like materials extends across a range of industrial, agricultural, and environmental sectors. Integrated environmental systems management means that the pressures on environmental compartments need to be regularly monitored and assessed. TE SQSs can also be used for this, and not just for assessing postindustrial development (http://www.aew.wur.nl/UK/Research/AQUATERRA/).

4.3 FRAMEWORKS FOR THE IMPLEMENTATION AND USE OF SQSs FOR TEs

Specific legislation at international, national or regional levels (e.g., the Water Framework Directive in Europe) exists for setting water quality criteria for the aquatic compartment. However, there is little such legislation for the soil compartment. In most countries, SQSs are provided by special legislation for contaminated sites. In some cases they are provided by soil and groundwater protection legislation and, in a few cases, by waste management legislation (Carlon 2007). In many countries, SQSs are defined by national legislation, but other countries leave this to the provincial or state authorities.

Many countries have SQSs for TEs. Historically, these elements have often been the only chemical contaminants for which assessments were made to determine potential environmental and human health risks. However, despite (or perhaps because of) this long history of monitoring and use of TE SQSs, there has been little development in the frameworks in which these standards are used. Some countries still use TE SQSs as single pass or fail criteria (Department of Environment [DoE]

1996) especially for the acceptance of materials applied to agricultural land (e.g., the Australian national (National Resource Management Ministers Council 2004) and state (e.g., EPA NSW 1997; EPA Victoria 2004) biosolids guidelines), while other countries have tiered approaches of assessment to deliver to a particular regulatory goal or driver, especially related to contaminated land (Jensen and Mesman 2006).

The use of a tiered system of assessment, with early tiers representing a generic precautionary screen, followed by increased levels of effort commensurate with increased levels of potential risk, is generally thought to be the most appropriate way to incorporate the use of SQSs and, according to Carlon (2007), is the most common approach in EU countries and Canada.

Following a desktop, site-specific "historical investigation," tier 1 would usually include a site investigation and some initial soil sampling (e.g., Environment Agency 2000). SQSs can be used at this tier to screen soils rapidly, exclude those for which there is little risk, and focus attention where potential risks are identified. A second tier may consist of in-depth sampling of any hot spots identified in the first tier investigation. This tier may also include the use of SQSs to delineate the spatial extent of contamination or the use of other risk assessment tools such as bioassays or in situ tests. The form and content of the tiers is legally and jurisdiction specific, but SQSs will often be tools and triggers for decisions in early tiers of the overall framework. Depending on the legal framework, a legal duty to carry out soil remediation is set either when soil concentrations are found that are above SQSs or when a risk assessment proves that the contamination poses a nonacceptable risk.

Australia uses an ecological risk-based process for assessing contaminated sites that has 3 tiers of investigation (NEPC 1999b). In tier 1, measured concentrations of a contaminant are compared to health investigation levels, interim urban ecological investigation levels (EILs) and groundwater investigation levels (NEPC 1999a). If measured concentrations exceed their corresponding investigation levels, then further investigation in the form of tier II is initiated. Tier II assessment is still largely a desktop study which determines site-specific investigation levels and reassesses the risk posed. In tier III, detailed site-specific assessments and extensive field measuring and monitoring are required. However, as the new proposed methodology for calculating SQSs (termed EILs in Australia) (Heemsbergen et al. 2009) aims to derive soil-specific SQSs, it is likely that the ecological risk assessment (ERA) process will be simplified to 2 tiers, by combining the previous tiers I and II (Warne, personal communication 2009).

Given the regulatory framework in which SQSs are used, regulators generally favor methods that are precautionary, particularly if they are used within a tiered risk-based framework and the extent of precaution decreases with each successive tier of investigation conducted, rather than underestimate risks and therefore potentially underprotect humans and soil ecosystems.

4.4 ACCOUNTING FOR AMBIENT BACKGROUND CONCENTRATIONS IN THE IMPLEMENTATION OF TE SQSs

TEs occur naturally and are present in the soil at concentrations that can vary quite considerably across different regions. For TEs the use of some environmental and human health risk assessment methodologies may derive generic limit values that are below ambient background levels (European Commission [EC] 2003). This can be due to several factors, including the following:

- The methodology accounts for data limitations through the application of precautionary assessment factors (e.g., division of the lowest no effect or effect concentration by a factor between 1 and 10), which results in lower SQSs.
- The soil background concentration of the TE spans orders of magnitude owing to the occurrence of local geological heterogeneity; populations may have developed tolerance to elevated concentrations, provided that the tolerance mechanisms are heritable.
- The laboratory test organisms from which effect concentrations are derived are generally cultured under conditions of high bioavailability or at low concentrations of TEs and therefore have a heightened potential susceptibility (International Council on Mining and Metals [ICMM] 2007).

There has been significant debate, particularly in Europe, about the appropriate definition of background concentrations. It is clear that deriving the "natural" background of a specific TE in soils exposed to more than 200 years of industrial development is difficult and may not be achievable. Therefore it is probably more appropriate to consider the "ambient background" concentration, i.e., the concentration of a metal in a soil that consists of both a natural geogenic fraction and an anthropogenic fraction (International Organization for Standardization [ISO] 2005). The anthropogenic fraction refers to moderate diffuse inputs into the soil and not inputs from local point sources that can result in appreciably elevated concentrations of TEs. The choice of whether an "ambient" or a geogenic background concentration is used is a decision that has to be taken by policy makers, accounting for the available scientific information and the regulatory context.

Having established a working definition for the background concentration (Section 2.2.1) in a soil, the second challenge is in actually deriving that concentration. The ABC can be determined by the selection of a relatively low percentile of concentrations (adjusted based on silt and organic matter concentrations) from a national soil monitoring database. It is used as a pragmatic, precautionary ABC (van Vlaardingen et al. 2005). Several studies have also attempted to develop relationships between soil physicochemical parameters, textural class, and soil metal concentrations (Hamon et al. 2004; Zhao et al., 2007). From these studies it is possible to estimate the ABC of a TE from another soil parameter (e.g., the Fe and Mn concentration). For some TEs, such as nickel and cobalt, these relationships are highly significant. However, for other elements such as zinc and lead, such relationships account for less than 50% of the variability, even when using relatively large data sets ($n > 4500$). Readers should refer to Chapter 2 on deriving SQSs for ecological receptors for greater detail on methods for determining ABC.

Having derived the ABC of a TE in a soil, there are 3 ways this information can be used in a regulatory framework to enable derivation of SQSs to support policy decisions. One way is practiced in the Netherlands for ecological SQSs and has been adopted by other countries and jurisdictions (e.g., Australia [Heemsbergen et al.

2008, 2009]) and is termed the "added risk approach" (Crommentuijn et al. 2000). The method of incorporating it into the SQSs is that the appropriate ABC is decided on, measured or estimated, and is added to a limit based on a normalized age corrected added contaminant concentration. The measured total soil concentration is then compared to the SQSs. A related concept is used in setting SQSs for human health where in some jurisdictions, human exposure via routes other than the contaminated soil, namely drinking water, consumer products, inhalation of air, and dietary consumption, are considered to be "background." The sum of these exposures is subtracted from the allowable exposure to a TE, leaving the residual as the basis for the SQS. This can result in SQSs below or close to background concentrations. Both of these concepts are further explored in the chapters describing ecological and human exposure.

The second method is termed the "total risk approach" and has been described by Euras (2007). The difference to the added risk approach is that the ABC and toxicity values calculated on an added basis (corrected for ageing) are summed, then normalized, and then the species sensitivity distribution analysis is conducted. This approach has been adopted in several of the EU risk assessments for metals, e.g., Cd, Cu, and Pb (Euras 2007).

In the third method, a "no risk approach" is used. In this the measured concentrations are compared directly to the ABC (Carlon 2007). This method therefore does not use toxicity-based SQSs. This method requires a geological survey of uncontaminated areas to establish background levels accurately (Reimann and Garrett 2005). However, there are numerous problems in determining geogenic and ambient background concentrations that weaken this method; for example, it is often not straightforward to delineate those soils that may have been influenced by anthropogenic activity in countries that may have had more than 250 years of industrial activity, although methods have been developed (Section 2.2.1, Chapter 2). Another limitation of this method is that background concentrations and those concentrations at which toxicity can occur often overlap (Section 2.2.1, Chapter 2). Therefore, using a high percentile (e.g., 90 percentile) of the background may lead to underprotection of soil ecosystems and humans that are not adapted to such high background levels. This problem could be overcome by using a considerably lower percentile of ABC, such as the 50th, but this may lead to an SQS that is much lower that the geogenic concentrations that are found in many parts of the country or region under consideration. Considering the problems with such SQSs, they should preferably be used as a trigger value in a tiered approach rather than as an action value. Which percentile should be used remains predominantly a policy decision, as the choice has major impacts on land management.

It is certainly not desirable nor scientifically defensible to have SQSs that are below the estimated ABC in any particular soil, region, or country. The use of SQSs that are below ABC is also a political decision, as the impact may be significant, especially if this would apply to large areas of land. It is important that such decisions are explicitly acknowledged and that the decision process is auditable and understandable by the regulated community. Nevertheless, if SQSs are set at or below ABC, then it is appropriate to revisit the conceptual site model drafted in the early tiers of the assessment and begin to involve relevant stakeholders to participate in redefining the problem. This would involve all exposure routes and prioritizing with the stakeholders where the main perceived and actual risks lie. This iteration and revision of the risk assessment will need to be communicated to the stakeholders in a managed and transparent way. However, the wisdom of setting an SQS for a TE that cannot be practically implemented is clearly questionable.

A recent review paper by Smolders et al. (2009) and the endorsed but not yet implemented Australian SQS derivation method (Heemsbergen et al. 2009) clearly demonstrate that it is possible to eliminate the setting of ecological-based SQSs below the ABC through a technically defensible derivation process. Importantly, this process is only viable for TEs for which there are significant ecotoxicity data and a well-developed understanding of fate and behavior. Of course, this is not the situation for many TEs that may have been identified by regulatory organizations as presenting a potential environmental or human health risk (van Vlaardingen et al. 2005).

Indeed, there is a dearth of guidance or understanding for the derivation and implementation of SQSs for TEs with limited relevant and reliable data (ICMM 2007). Therefore some countries have not implemented SQSs for TEs with limited data sets, which result in high residual uncertainty and for which there is poor understanding of the range of soil background concentrations (Environment Agency 2008a), while others have made political decisions to enable implementation while broadly acknowledging the inherent uncertainty in use (van Vlaardingen et al. 2005).

4.5 ACCOUNTING FOR (BIO)AVAILABILITY IN THE DERIVATION OF TE SQSs

The default assumption in risk assessment is generally to assume that a contaminant in soil is 100% bioavailable, which is rarely if ever true. One key factor hindering the usefulness of ecological and human health SQSs based on total TE concentrations in soil is that bioavailability is not always properly considered. It has long been known, qualitatively, that soil physicochemical properties affect the toxicity and bioavailability of TEs and other contaminants (Alloway 1995; Basta et al. 2005). The lack of robust, nonoperationally defined, quantitative relationships between soil physicochemical properties and bioavailability has until very recently meant that SQSs can only be derived by 2 main methods:

- The first and more conservative approach is to derive SQSs that protect receptors under worst case scenarios (e.g., those where the bioavailability is assumed to be 100%), using total concentration data. This approach automatically means that receptors will be overprotected in most soils.
- The second approach is to acknowledge that soil properties affect bioavailability and to incorporate this into the derivation of the SQSs. In this approach, soils are divided into either 2 (CCME 1999, 2006) or 3 characteristic classes (BBodSchV 1999) and separate SQSs are derived using toxicity data (expressed as total concentrations) generated in soils within each soil class. This concept of subdividing soils based on the potential bioavailability of contaminants in the soil is also part of the USEPA method of deriving Ecological Soil Screening Levels (USEPA 1996). It should be noted that this approach does not take the effect of mineralogy of the TE on bioavailability into account.

It is common practice to define a TE SQS as a "total" (concentrated acid soluble) concentration. The environmental or human health relevance of this metric is, however, questionable and has been routinely demonstrated to be overprotective (e.g., Nolan et al. 2009). Numerous soil extractions or speciation methods have been proposed to measure so-called "bioavailable fractions" (McLaughlin et al. 2000). In some jurisdictions an operationally defined soluble fraction of metals, such as DTPA, EDTA, or 1M NH_4NO_3, have been or are still used (Interdepartmental Committee on the Redevelopment of Contaminated Land 1990; Swiss Agency for the Environment, Forestry and Landscape 2001; BBodSchV 1999; ISO 2008). In theory, if toxicity data were expressed in terms of a perfect measure of bioavailability, then the toxicity for a TE to a single species would be identical in all soils (McLaughlin et al. 2000). A number of studies have tested this argument and compared the variation in toxicity of TEs in a variety of soils. These comparisons have found that the supposed chemical measures of bioavailability do not reduce the unexplained variation in toxicity as compared to total concentrations (Smolders et al. 2003, 2004; Oorts et al. 2006; Zhao et al. 2006; Broos et al. 2007; Warne et al. 2008a). Given these findings, there is no generally accepted method within the scientific community for predicting TE toxicity in soils. A "pseudototal concentration" defined, for example, through the use of aqua regia digestion, is favored by most regulators as it is precautionary, simple, and cheap to undertake; well standardized; can be measured reasonably reliably; and is relatively easy to understand by a range of stakeholders who may not be technical experts (Umweltbundesamt [UBA] 2008).

In the last decade, considerable effort has been made to demonstrate how metal bioavailability to the terrestrial ecosystem is affected by the abiotic properties of soil. Bioavailability models have been developed using data from more than 500 new chronic toxicity tests in a wide range of European soil types. These models were applied to several metals in EU risk assessments required under the Existing Substances Regulation 793/93EC. Smolders et al. (2009) collated this information and showed how these data have been used in the EU for defining ecologically based safe values for cadmium, copper, cobalt, nickel, lead, and zinc in soil. Similar work has been conducted in Australia, with relationships estimated between soil properties and bioavailability to wheat plus 2 measures of soil microbial health (substrate induced nitrification and respiration) for copper and zinc (Broos et al. 2007; Warne et al. 2008a, 2008b) and for plant uptake of cadmium (McLaughlin et al. 2006). A limitation of these models is that they are empirical rather than mechanistic, but if the predictions are robust, accurate, and practical, this may be more of an academic question than one of importance for regulatory implementation.

Recent guidance on the incorporation of bioavailability-based concepts into regulation has generated considerable discussion among EU regulators (MERAG and HERAG) (ICMM 2007). Some believe that the evidence presented in this and other documents (Netherlands 2004; Denmark 2008) do not describe the mechanistic processes sufficiently to allow bioavailability to be accounted for in regulatory frameworks (UBA 2008). In contrast, some EU Member States have taken the view that these concepts and their regulatory adoption do represent a significant advance in existing methods for assessing TE risks in soils that have in the past been largely based on so-called "expert judgment" (Environment Agency 2008a; Vlaams Reglement

Bodemsanering [VLAREBO] 2008). Australia is also currently considering adopting biosolids quality guidelines (Heemsbergen et al. 2009) for cadmium, copper, and zinc, which incorporate bioavailability models into the derivation methodology. These values have subsequently been incorporated into a method currently being considered for adoption by Australia to derive fertilizer contaminant upper limits (Sovari et al. 2009), and these bioavailability models have been the endorsed but not yet implemented Australian SQS derivation method (Heemsbergen et al. 2009).

In order to assess the validity of bioavailability-based approaches to the assessment of TE risk in soils the Environment Agency (2008b) of England and Wales tested the predicted no effect concentrations and accompanying bioavailability assessments under the Existing Substance Regulations (793/93/EEC) for cadmium, copper, lead, nickel, and zinc (Netherlands 2004; Belgium 2007; European Copper Institute 2007; Lead Development Association International 2007; Denmark 2008). This validation exercise was completed through a comparison with the existing and widely used UK Sludge Use in Agriculture Regulations (HMSO 1989), UK Sewage Sludge Code of Practice (DoE 1986), and the EC 3rd Working Document on Sludge (EC 2000). The validation was possible because of the existence of matched metal concentration (including controls) and biological ecotoxicity test data (including microbial biomass, respiration rate, rhizobia numbers, and yield) for over 500 soil samples. Of these 500 soil samples, a subset of 100 soils varying in land use, management, and soil characteristics was selected to represent the widest range of metal exposure conditions. Four performance categories were used to assess how well the sets of SQSs predicted the observed toxic effects:

- Pass/pass: the soil concentration was below the metal limit value for the respective regime and no significant biological effect was observed.
- Fail/fail: the soil concentration was above the metal limit value for the respective regime and a significant adverse biological effect was observed.
- Pass/fail: the soil concentration was below the metal limit value, but adverse effects on soil biology were observed.
- Fail/pass: the soil concentration was above the metal limit value but no adverse biological effects were observed.

For the soils that belonged to the pass/pass or fail/fail categories the SQSs were considered to provide an appropriate level of protection. In contrast, for the soils that belonged to the pass/fail and the fail/pass categories the SQSs were considered to be either under- or overprotective. A summary of the results from this assessment are shown in Figure 4.1. It can clearly be seen that the current UK regimes (the first 2 bars in Figure 4.1) are underprotective for between 25 and 34% of the soils tested, whereas the EU Sludge Working Document limits and the bioavailability-based limits derived in the EU risk assessments both perform considerably better, being underprotective for only ≈6% of the soils. However, the EU Sludge Working Document limits and the bioavailability-based limits are overprotective in twice the number of cases (i.e., 17% of soils) when compared to use of the existing UK limits (8 to 9%). In total the percentage of soils that are either under- or overprotected by the EU Sludge Working Document limits and EU risk assessments (23 to 24%) is

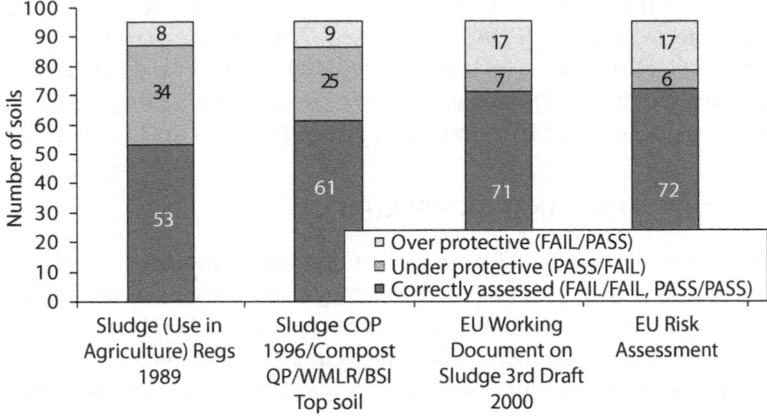

FIGURE 4.1 Number of soils for which each regime was over, under or sufficiently protective.

still markedly lower than that when the current UK guidelines are applied (34 to 42%). This analysis clearly shows that the bioavailability-based SQS provides a more appropriate degree of environmental protection.

4.6 ACCOUNTING FOR MIXTURES OF TEs IN REGULATORY FRAMEWORKS

Mixture effects are the combined toxicity caused by the simultaneous presence of TEs in soil. Such effects can be additive (the total toxicity equals the sum of the toxicities of individual elements, which must have the same mode of action) or synergistic or antagonistic (when one element enhances or reduces the toxicity of another element). At this time, there are no jurisdictions where mixtures of metals at a site are dealt with any other way than simply regulation based on the TE that most exceeds the SQSs. The reasons for this are that there are simply too few data to meaningfully inform how regulations should be modified. Research work is ongoing, but results are neither sufficiently robust nor have they been sufficiently validated. The exception to this is the tBLM, which accounts for antagonism with major cations in the uptake of TEs.

There is an obvious need for information on combined toxicity of TEs to humans and biota. It is possible when assessing ecological risks to estimate mixture toxicity using the potentially affected fraction for multiple substances (ms-PAF) (details of this method are provided by Posthuma et al. (2008).

For ecosystems, using toxicity testing protocols such as direct toxicity assessment and whole effluent toxicity testing are a viable, in fact, preferable alternative to chemical measures of risk assessment, but these are generally most valuable in latter tiers of a tiered approach. These tests directly provide information on the toxicity of mixtures in the soil on the test species irrespective of the type of chemical interaction involved. However, depending on the question being addressed, they can be difficult to interpret, i.e., what specific contaminant or mixture of contaminants is causing the toxicity? If it is crucial to determine which TE constituent in a mixture is contributing to the toxicity, then the toxicity identification evaluation (TIE) framework (also referred to as effect directed analysis [EDA]) can be used. While this framework is well developed for aquatic ecosystems (e.g. USEPA 1991, 1993a, 1993b) and has recently been developed for sediments (Phillips et al. 2009; Weston et al. 2009), relatively little work on the application of TIE to terrestrial and soil systems has been conducted.

4.7 MONITORING AND ASSESSMENT

As mentioned in Section 4.3, SQSs are often used as concentration levels to trigger further action. Soil quality can be determined by contaminant concentrations in soil and groundwater; those concentrations may be compared to the SQSs or may be used as input in risk assessment models. Using measured soil concentrations in this way is a very objective framework for policy decisions as the outcome of this exercise is a very strict one, and there may be uncertainty in contaminant concentration. Soil and groundwater sampling as well as chemical analyses are essential steps to obtain soil quality data to be compared to the SQSs. Contaminant concentrations obtained from sampled soils will vary in the horizontal and vertical directions and errors in sampling, sample preparation and analysis will add significantly to uncertainty in soil data.

Most countries have at least general guidance documents on sampling and analysis that assist in reducing uncertainty in the risk assessment owing to spatial variability in TE concentration. Some countries have more detailed guidelines that state how samples should be collected, the minimum number of samples, the sampling strategy to be adopted, TEs to be analyzed, and the analytical methods. There is also guidance readily available on the sampling of piles and heaps of soil-like material (e.g., BS EN 2000).

Importantly, if efforts to implement bioavailability-based TE approaches into regulatory frameworks are to be successful, then it is important to address the requirement for additional supporting physicochemical parameters in order to facilitate that implementation. These additional data, such as pH, organic carbon, cation exchange capacity, and clay content, provide contextual background to interpret the measured TE concentrations and also to allow the application of empirical relationships that can deliver robust assessments of potential TE ecological and human health risks (Van Gestel 2007; Environment Agency 2008a).

4.8 DATA SOURCES

Carlon (2007) noted that national authorities within the EU use different (eco) toxicological databases. There are several large and comprehensive ecotoxicology databases such as the USEPA ECOTOX database (USEPA 2007b) and the RIVM e-TOXBASE (http://www.e-toxbase.com/), which aim to collate data from around the world. There are also smaller databases that contain toxicity data only for specific geographical regions such as the Australasian Ecotoxicology Database (Warne et al. 1998; Warne and Westbury 1999; Markich et al. 2002). Considerable effort is expended in establishing and maintaining these databases, but, unfortunately, there is also considerable duplication of the data they hold.

A vital part of deriving SQSs is to assess the quality of the toxicity data, because as with any modeling exercise the quality of the data to a large degree controls the quality of the outputs. Again, different assessment methods are used by the different databases, although the system used in the Australasian Ecotoxicology Database is modeled on that used in the USEPA ECOTOX database.

While different types of toxicity data may be used by different jurisdictions to derive SQSs (e.g., some may use EC10 data while others only use NOEC data), it should be possible to assess the quality of the data using one universally agreed upon methodology. The assessment of data quality is a vital step in deriving SQSs, and it should therefore be conducted by people with appropriate expertise.

Similarly, there is a clear need for a common database on exposure data and information related to exposure assessment, e.g., food habits, TE content in food items, bioconcentration factor values, and the algorithms used to derive SQSs. While the use of exposure data is highly dependent on the geographical and climatic conditions and sociology, the availability of a common pool of such data would ensure that relevant data of sufficient quality are used. Such a database would preferably be hosted and maintained by an international agency. The same logic equally applies to hosting an effects (ecotoxicology data) database.

A thorough international review of the individual models and the algorithms would be beneficial; it should have as minimal goal to show the boundary conditions of algorithms (making it easier for countries to choose algorithms fit for their conditions). It could also be a trigger for a harmonization of similar modules, making the outcomes of different models comparable.

4.9 COMMUNICATION

It is important to involve stakeholders in the SQS implementation process, preferably at an early in that process. While such involvement requires more resources, good planning, and skilled discussion through facilitators, it is likely to be beneficial in the longer term (e.g., higher acceptance and more efficient implementation).

Nonacceptance of SQSs by stakeholders is often driven by uncertainty caused by a lack of understanding and absence of clear communication. Indeed, risk assessments and investigation reports are complex. There is a clear need for scientists, consultants, and regulators to translate the results in a clear and transparent way, providing clarifications on the meaning of different types of SQSs, what exceedance means in practice. Keys for a successful communication include timely communication in a transparent form that is sympathetic to the concerns of the affected parties or stakeholders. Information, handbooks, and examples on how to manage risk communication are readily available from many disciplines, including food safety (http://www.fao.org/docrep/005/X1271E/X1271E03.htm), flooding (http://www.defra.gov.uk/environment/flooding/documents/manage/shaldon.pdf), and contaminated land (http://www.health.vic.gov.au/environment/downloads/hs680_land_contam.pdf).

Communication between scientists and regulators can be a source of potential misunderstandings for several reasons, including the following:

- lack of clear objectives set by regulators;
- overexpectations by regulators of science solutions to practical problems;

- lack of simple, easy to understand conclusions and their practical implications by scientists;
- lack of guidance on how to handle uncertainty; and
- limited practical understanding by scientists of implementation challenges.

Often too little discussion takes place between scientists and regulators to agree on what science is able to realistically deliver at a specific moment in time and how the results of science can best be implemented.

4.10 CONCLUSIONS

Knowledge and understanding of the fate and toxicity of TEs in soils have progressed significantly in the past 10 years. However, despite this progress, especially in assessing potential human health and environmental risks, uptake, and integration of this knowledge into regulatory frameworks internationally is patchy at best. The reasons for this are mixed but are likely to include the view that the estimation of soil ingestion rates, the identification of appropriate values for background exposure, and the science underpinning bioavailability correction are still only partly developed or that some of the approaches may not be sufficiently precautionary. Yet there are some jurisdictions in which the reliance of traditional methods of soil standard setting have clearly been shown to be underprotective (Section 4.5)

Nevertheless, recently some countries have attempted to assess the form and magnitude of this apparent uncertainty of the more advanced methods through comparison with existing systems that are based on older and less well developed science. These showed the higher accuracy of the more advanced methods and the lower chance of giving false results (especially less underprotective). Furthermore, there are successful examples of where the organizational and logistical challenges associated with implementing robust evidence-based quality standards have been shown to be significantly less onerous than some have suggested (UBA 2008). It is now necessary for technical specialists to bridge the divide that exists between themselves and policy makers in order to facilitate the adoption of fit for purpose "evidence-based" systems of TE risk assessment and regulation.

REFERENCES

Alloway BJ. 1995. Heavy metals in soils. 2nd ed. London (UK): Blackie.

Arshad MA, Martin S. 2002. Identifying critical limits for soil quality indicators in agroecosystems. Agricult Ecosys Environ 88:153–160.

Basta NT. Ryan JA, Chaney RL. 2005. Trace element chemistry in residual-treated soil: key concepts and metal bioavailability. J Environ Qual 34:49–63.

BBodSchV. 1999. Bundes-Bodenschutz- und Altlastenverordnung (BBodSchV) vom 12. Juli 1999. (Federal Soil Protection and Contaminated Sites Ordinance dated 12 July 1999). Bundesgesetzblatt I, 1554.

Belgium. 2007. European Union risk assessment report on cadmium oxide and cadmium metal, Belgium, 2005. Institute for Health and Consumer Protection European Chemicals Bureau. Report prepared by Belgium, Federal Public Service Health, Food Chain Safety and Environment on behalf of the European Union.

Brandes LJ, Den Hollander HA, Van de Meent D. 1996. SimpleBox 2.0: A nested multimedia fate model for evaluating the environmental fate of chemicals. Report 719101029. Bilthoven (Netherlands): National Institute for Public Health and the Environment (RIVM).

British Standards Institution (BSI). 2007. Specification for topsoil BS 3882:2007. London (UK): British Standards Institution.

Broos K, Warne M, Heemsbergen DA, Stevens D, Barnes MB, Correll RL, McLaughlin MJ. 2007. Soil factors controlling the toxicity of Cu and Zn to microbial processes in Australian soils. Environ Toxicol Chem 26(4):583–590.

BS EN. 2000. Soil improvers and growing media. Sampling. Report BS EN 12579:2000. London (UK): BSI British Standards.

Canadian Council for Ministers for the Environment (CCME). 1999. Canadian environmental quality guidelines. http://www.ccme.ca/publications/list-publications.html. Accessed on 30 Sept. 2009.

Canadian Council for Ministers for the Environment (CCME). 2006. A protocol for the derivation of environmental and human health soil quality guidelines. Winnipeg (Canada): CCME.

Carlon C. 2007. Derivation methods of soil screening values in Europe: a review and evaluation of national procedures towards harmonisation. Report EU 22805-EN. Ispra (Italy) : European Commission, Joint Research Centre. 306 p.

Crane M, Matthiessen P, Stretton Maycock D, Merrington G, Whitehouse P. 2009. Derivation and use of environmental quality and human health standards for chemical substances in water and soil. Pensacola (FL): Society of Environmental Toxicology and Chemistry (SETAC).

Crommentuijn T, Polder M, Sijm D, De Bruijn J, Van De Plassche E. 2000. Evaluation of the Dutch environmental risk limits for metals by application of the added risk approach. Environ Toxicol Chem 19:1692–1701.

Denmark. 2008. European Union risk assessment report on nickel, nickel sulphate, nickel carbonate, nickel chloride, nickel dinitrate, Denmark. Report prepared by Denmark, Danish Environmental Protection Agency on behalf of the European Union.

Department of Environment (DoE). 1986. Code of practice for agriculture use of sewage sludge. London (UK): Department of the Environment.

Environment Agency. 2000. Technical aspects of site investigation volume I (of II) overview. Technical Report P5-065/TR. Bristol (UK): Environment Agency. Also available at http://www.environment-agency.gov.uk/static/documents/3-SP5-065-TR-e-e.pdf.

Environment Agency. 2008a. Guidance on the use of soil screening values in ecological risk assessment. Science Report SC070009/SR2b. Bristol (UK): Environment Agency. Also available at http://publications.environment-agency.gov.uk/pdf/SCHO1008BOSQ-e-e.pdf.

Environment Agency. 2008b. Road testing 'trigger values' for assessing site specific soil quality. Phase 1 — Metals. Science Report SC050054SR1. Bristol (UK): Environment Agency.

EPA NSW. 1997. Environmental guidelines: use and disposal of biosolids products. Sydney (Australia): EPA NSW.

EPA Victoria, 2004. Guidelines for environmental management. Biosolids land application. Publ. 943, Melbourne (Australia): EPA Victoria. 74 p.

Euras. 2007. Fact sheet 1: risk characterization, general aspects. MERAG Program — building block, risk assessment. Annex 1, added vs total risk approach and its use in risk assessment and/or environmental quality setting. ICMM, UK. http://www.icmm.com/page/1185/metals-environmental-risk-assessment-guidance-merag (accessed 24 September 2009).

European Commission (EC). 2000. Working Document on Sludge Third Draft ENV.E.3/LM April 2000.

European Commission (EC). 2003. Technical Guidance Document on Risk Assessment in support of Commission Directive 93/67/EEC on Risk Assessment for new notified substances, Commission Regulation (EC) No 1488/94 on Risk Assessment of existing substances and Directive 98/8/EC of the European Parliament and of the Council concerning the placing of biocidal products on the market. European Commission, Joint Research Centre, EUR 20418 EN.

European Copper Institute (ECI). 2007. European Union risk assessment report on copper, copper(ii)sulphate pentahydrate, copper(i)oxide, copper(ii)oxide, dicopper chloride trihydroxide. Voluntary risk assessment, draft February 2007. Brussels (Belgium): ECI.

Hamon RE, McLaughlin MJ, Gilkes RJ. 2004. Geochemical indices allow estimation of heavy metal background concentrations in soils. Global Biogeochem Cycles 18(GB1014). doi:10.1029/2003GB002063.

Heemsbergen D, Warne M, McLaughlin M, Kookana R. 2008. Proposal for an Australian methodology to derive ecological investigation levels in contaminated soils. Science Report 15/08. Adelaide (Australia): CSIRO Land and Water, 77p.

Heemsbergen DA, Warne M, Broos K, Bell M, Nash D, McLaughlin MJ, Whatmuff M, Barry G, Pritchard D, Penney N. 2009. Application of phytotoxicity data to a new Australian soil quality guideline framework for biosolids. Sci Total Environ 407:2546–2556.

HMSO. 1989. Statutory Instrument. The sludge (use in agriculture) Regulations 1989, SI1263. London (UK): HMSO.

Interdepartmental Committee on the Redevelopment of Contaminated Land (ICRCL). 1990. Notes on the restoration and aftercare of metalliferous mining sites for pasture and grazing. Guidance Note 70/90 1st ed. London (UK): Department of Environment.

International Council on Mining and Metals (ICMM). 2007. Metals environmental risk assessment guidance. London (UK): ICMM.

International Organization for Standardization (ISO). 2005. Soil quality: guidance on the determination of background values. Geneva (Switzerland): ISO. (ISO CD 19258).

International Organization for Standardization (ISO). 2008. Soil quality: requirements and guidance for the selection and application of methods for the assessment of bioavailability of contaminants in soil and soil materials. Geneva (Switzerland): ISO. (ISO 17402:2008).

Jensen J, Mesman M. 2006. ecological risk assessment of contaminated land; decision support for site specific investigations. Report 711701047. Bilthoven (Netherlands): RIVM. 136 p.

Lead Development Association International (LDAI). 2007. European Union risk assessment report on lead metal, lead oxide, lead tetraoxide, lead stabiliser compounds. Voluntary risk assessment, draft April 2007. London (UK): LDAI.

Mackay D. 2001. Multimedia environmental models: the (fugacity approach. Chelsea (MI): Lewis Publishers.

Markich SJ, Warne M, Westbury A-M, Roberts CJ. 2002. A compilation of toxicity data for chemicals to Australasian species. Part 3: Metals and metalloids. Aust J Ecotoxicol 8(1):1–137.

McLaughlin MJ, Whatmuff M, Warne M, Heemsbergen DA, Barry G, Bell M, Nash D, Pritchard D. 2006. A field investigation of solubility and food chain accumulation of biosolid-cadmium across diverse soil types. Environ Chem 3:428–432.

McLaughlin MJ, Zarcinas BA, Stevens DP, Cook N. 2000. Soil testing for heavy metals. Comm Soil Sci Plant Anal 31:1661–1700.

Merrington G, Boekhold S, Haro M-A, Knauer K, Romijn K, Sawatsky N, Schoeters I, Stevens R, Swartjes F. 2009. Derivation and use of environmental quality and human health standards for chemical substances in groundwater and soil. In: Crane M, Matthiessen P, Stretton Maycock D, Merrington G, Whitehouse P, editors. Derivation and use of environmental quality and human health standards for chemical substances in water and soil. Boca Raton (FL): SETAC. p 105–126.

Ministry for the Environment. 2003. Contaminated land management guidelines no. 2. Wellington (New Zealand): Ministry for the Environment.

National Environment Protection Council (NEPC). 1999a. National Environment Protection (assessment of site contamination) Measure 1999. Schedule B(1) Guideline on the investigation levels for soil and groundwater. Adelaide Australia): NEPC. 16 p.

National Environment Protection Council (NEPC). 1999b. National Environment Protection (Assessment of Site Contamination) Measure 1999. Schedule B(5) Guideline on Ecological Risk Assessment. NEPC, Adelaide, Australia. 50p.

National Resource Management Ministers Council (NRMMC). 2004. Guidelines for sewerage systems — biosolids management. Strategy Paper 13. Canberra (Australia): NRMMC. 45 p.

Netherlands. 2004. European Union risk assessment report on zinc metal, zinc(ii)chloride, zinc sulphate, zinc distearate, zinc oxide, trizinc bis(orthophosphate). Prepared by the Netherlands, RIVM on behalf of the European Union.

Nolan AL, Ma Y, Lombi E, McLaughlin MJ. 2009. Speciation and isotopic exchangeability of nickel in soil solution. J Environ Qual 38:485–492.

Oorts K, Ghesquiere U, Swinnen K, Smolders E. 2006. Soil properties affecting the toxicity of $CuCl_2$ and $NiCl_2$ for soil microbial processes in freshly spiked soils. Environ Toxicol Chem 25:836–844.

Phillips BM, Anderson BS, Hunt JW, Clark SL, Voorhees JP, Tjeerdema RS, Casteline J, Stewart M. 2009. Evaluation of phase II toxicity identification evaluation methods for freshwater whole sediments and interstitial water. Chemosphere 74:648–653.

Posthuma L, Richards S, De Zwart D, Dyer S, Sibley P, Hickey C, Altenberger R. 2008. Mixture extrapolation approaches. In: Solomon K, Brock T, de Zwart D, Dyer S, Posthuma L, Richards S, Sanderson H, Sibley P, Van den Brink P, editors. Extrapolation practice for ecotoxicological effect characterization of chemicals. Boca Raton (FL): Taylor & Francis. p 137–185.

Reimann C, Garrett RG. 2005. Geochemical background – concept and reality. Sci.1269 Total Environ. 350, 12–27.

Royal Commission on Environmental Pollution (RCEP). 1998. 21st report: setting environmental standards. London (UK): RCEP.

Smolders E, Buekers J, Oliver I, McLaughlin MJ. 2004. Soil properties affecting toxicity of zinc to soil microbial properties in laboratory-spiked and field-contaminated soils. Environ Toxicol Chem 23:2633–2640.

Smolders E, Buekers J, Waegeneers N, Oliver I, McLaughlin MJ. 2003. Effects of field and laboratory Zn contamination on soil microbial processes and plant growth. Final report to the International Lead and Zinc Research Organisation. Leuven (Belgium): Katholieke Universiteit Leuven and CSIRO. 67 p.

Smolders E, Oorts K, Van Sprang P, Schoeters I, Janssen CR, McGrath S, McLaughlin M. 2009. The toxicity of trace metals in soil as affected by soil type and ageing after contamination: using calibrated bioavailability models to set ecological soil standards. Environ Toxicol Chem 28(8):1633–1642.

Sorvari J, Warne M, McLaughlin MJ, Kookana R. 2008. Methodology to assess the environmental impacts of mineral fertilisers, including industrial residues and to derive contaminant limits. CLW Science Report prepared for the Primary Industries Ministerial Council and the Environment Protection and Heritage Council. 75 p.

Sorvari J, Warne M, McLaughlin MJ, Kookana R. 2009. Investigation into the impacts of contaminants in mineral fertilisers, fertiliser ingredients and industrial residues and the derivation of guidelines for contaminants. CLW Science Report 25/09 prepared for the Primary Industries Ministerial Council and the Environment Protection and Heritage Council. 102 p.

Sparling G, Lilburne L, Vojvodic-Vukovic M. 2003. Provisional targets for soil quality indicators in New Zealand. Lincoln (New Zealand): Landcare Research New Zealand Ltd.

Swiss Agency for the Environment, Forestry and Landscape. 2001. Commentary on the Ordinance 1 July 1998 relating to impacts on the soil. Berne (Switzerland): Swiss Agency for the Environment, Forestry and Landscape. 43 p.

Umweltbundesamt (UBA). 2008. Incorporation of metals bioavailability into regulatory frameworks. Report UBA-FG IV 2.3. Dessau (Germany): UBA.

USEPA. 1991. Methods for aquatic toxicity identification evaluations: phase I toxicity characterisation procedures. 2nd ed. Washington (DC): USEPA. EPA/600/6-91/003.

USEPA. 1993a. Methods for aquatic toxicity identification evaluations: phase II toxicity identification procedures for samples exhibiting acute and chronic toxicity. Report EPA/600/R-92/080. Washington (DC): USEPA.

USEPA. 1993b. Methods for aquatic toxicity identification evaluations: phase III toxicity confirmation procedures for samples exhibiting acute and chronic toxicity. Report EPA/600/R-92/081. Washington (DC): USEPA.

USEPA. 1996. Soil screening guidance: technical background. Report EPA/540/R-95/128. Washington (DC): USEPA. Also available http://www.epa.gov/ecotox/ecossl/.

USEPA. 2007a. Framework for metals risk assessment. Report EPA 120/R-07/001. Washington (DC): USEPA. Also available at http://www.epa.gov/osa/metalsframework/pdfs/metals-risk-assessment-final.pdf.

USEPA. 2007b. ECOTOX user guide: ECOTOXicology Database System. Ver. 4.0. http:/www.epa.gov/ecotox/. Accessed August 2008.

Van Gestel G. 2007. Setting new soil standards for metals: improved relations with soil properties. Poster at SETAC Europe 17th Meeting, Multiple Stressors for the Environment and Human Health Present and Future Challenges and Perspectives. 20–24 May. Porto (Portugal).

van Vlaardingen PLA, Posthumus R. Posthuma-Doodeman CJAM. 2005 Environmental risk limits for nine TEs. Report 601501029. Bilthoven (Netherlands): RIVM.

Vlaams Reglement Bodemsanering (VLAREBO). 2008. Implementation order of the Flemish soil decree ratified 14 December 2007. Published 22 April 2008.

Warne M, Heemsbergen DA, McLaughlin MJ, Bell M, Broos K, Whatmuff M, Barry G, Nash D, Pritchard D, Penney N. 2008a. Models for the field-based toxicity of copper and zinc salts to wheat in eleven Australian soils and comparison to laboratory-based models. Environ Pollut 156(3):707–714.

Warne M, et al. 2008b. Modelling the toxicity of Cu and Zn salts to wheat in fourteen soils. Environ Toxicol Chem 27:786–792.

Warne M, Westbury A-M. 1999. A compilation of toxicity data for chemicals to Australasian species: Part II. Organic chemicals. Aust J Ecotoxicol 5:21–85.

Warne M, Westbury A-M, Sunderam R. 1998. A compilation of toxicity data for chemicals to Australasian aquatic species: Part 1. Pesticides. Aust J Ecotoxicol 4:93–144.

Weston DP, You J, Harwood AD, Lydy MJ. 2009. Whole sediment toxicity identification evaluation tools for pyrethroid insecticides: III. Temperature manipulation. Environ Toxicol Chem 28(1):173–180.

Zhao FJ, McGrath SP, Merrington G. 2007. Estimates of ambient background concentrations of trace metals in soils for risk assessment. Environ Pollut 148:221–229.

Zhao FJ, Rooney CP, Zhang H, McGrath SP. 2006. Comparison of soil solution speciation and diffusive gradients in thin films measurement as an indicator of copper bioavailability to plants. Environ Toxicol Chem 25:733–742.

5 Recommendations for the Derivation of Interpretable and Implementable Soil Quality Standards for Trace Elements

Graham Merrington, Ilse Schoeters, Michael Warne, Beverley Hale, and Mike J. McLaughlin

5.1 INTRODUCTION

Trace elements (TEs) present a range of regulatory challenges that have received significant attention over the past 5 to 10 years stimulated by research programs borne largely from the EU Existing Substance Regulations (793/93/EEC). Advances in the understanding of the behavior and fate of TEs in soils in terms of potential human and environmental health are leading to a reassessment of existing methods of how risks posed by TEs are assessed.

The previous chapters have provided an indication of the current status of much of this research and how this has been developed by researchers and adopted by individual jurisdictions. There is significant overlap in the previous chapters in regard to several key issues that routinely differ among jurisdictions and for which the technical understanding is not clear-cut. The use of ambient background concentrations (ABCs) of TEs with any standard setting regime is such an example. However, there is clearly a greater level of agreement on issues such as accounting for TE bio(availability), for which most jurisdictions recognize a need, but for which there is perhaps a shortfall in consistent processes and tools to pragmatically implement. Models play a significant role in underpinning the derivation of ecological and human health standards. Data input quality and validation are often key factors in development and use, but there should be little that is contentious in terms of underlying principles, and therefore consensus and harmonization of these processes should be possible. Perhaps the regulatory adoption and use of chronic aquatic Biotic Ligand Models (BLMs) by several member states in Europe provides a useful example of how progress in this area may be attained (Environment Agency 2009).

However, it should be stressed, and is perhaps of little surprise, that this journey has not always been a smooth one (Umweltbundesamt 2008).

The following section provides several recommendations of what we consider to be key needs in this area and illustrates some best practices in delivering robust, interpretable, and implementable soil quantity standards (SQSs) for TEs. These practices are the recommendations from the individual chapters but should be viewed as a single entity in order to fulfill the need for a consistency of protection across receptors.

5.2 SOIL QUALITY STANDARDS FOR TEs AND BEST PRACTICE

The TEs for which we have the most developed understanding and subsequently have the most interpretable (in terms of potential risks) SQSs are those for which we have the most effect and exposure data. However, there is a need to continue to develop and improve the relationships between ecotoxicological effects of TEs and soil composition by performing experimental work to extend the ranges of soil types and the TEs for which empirical observations are available. This is especially the case for TE oxyanions. There is also a need for validation of empirical relationships across species and taxa, and validation of laboratory-based relationships in field-contaminated soils, or to develop and validate field-based relationships as this would improve confidence in the widespread use of normalization relationships for SQS setting.

There is a research need to improve the mechanistic underpinning of the BLM for terrestrial organisms by identifying 1) sites of TE uptake and mechanisms of toxicity and 2) measures of organism TE content that correlate with toxic effects and can be used as proxies for biotic ligand-bound TE. Conflicting results for effects of competing ions on TE toxicity under controlled conditions need to be resolved.

A range of selective extraction techniques have been, and continue to be, used to assess potential risks from TEs in soils (and wastes). There is a research requirement to assess whether these methods are in fact better measures of the bioavailable fraction than total concentrations, in conjunction with modeling studies and using the same data sets to allow direct comparisons of methods.

Generally, the most overlooked facet of standard setting is implementation. Examples can be seen throughout the previous chapters where standards have been derived using methods not designed for use on TEs that deliver values below ABCs or that would result in deficiencies of essential TEs in biological systems. Therefore tiered approaches to implementation and use of TE SQSs are recommended as these represent the most effective, proportionate, and efficient way of assessing potential risks in the environment.

Different SQS derivation models are used by different countries (Carlon 2007). While geographical and sociocultural reasons can drive differences among SQSs, it is often unclear why certain data or algorithms are selected. Indeed, such reasoning can be lost in time. The foundations on which these relationships are built need to be traced and reviewed on a regular basis and not just uncritically accepted as the "gospel." A thorough international review of the individual models and their algorithms is needed to clarify differences, the supporting rationale of the differences, and the application domains of the models — this would make the selection of the appropriate models and algorithms easier. Such a role would preferably be conducted by an international agency.

Empirical models that can predict bioavailability and toxicity based on soil properties have been developed and used to derive soil-specific ecologically based SQSs. Their use can reduce the uncertainty associated with SQSs. These models have been validated with data from field experiments in the United Kingdom and are being applied in the European Union risk assessments, in Australia, and in Flanders. However, it is imperative that scientists and researchers understand the regulatory context in which these models are going to be used. This regulatory context may represent a potentially uncomfortable practical compromise for researchers between the scientific foundations of the model and regulatory needs. These issues require careful consideration.

Collating and assessing the quality of ecotoxicology data for TEs and all other contaminants is a large undertaking and there is currently considerable duplication of effort (e.g., ECOTOX, RIVM database, ECOTOC, and the Australasian Ecotoxicology Database). We recommend that a single international database with a single (but adjustable) method of assessing the quality of the data be established. This should also include the harmonization of criteria used for screening the relevance and reliability of nonstandardized soil toxicity tests in order to increase consistency (e.g. Registration, Evaluation, Authorisation and restriction of Chemicals in Europe [REACH] and Globally Harmonised Systems [GHS]). It would be preferable if an international organization had responsibility for this.

Finally, we recommend that greater use be made of field data in the derivation of SQSs and the development of regulatory frameworks for implementation and use of TEs in soils (De Jong et al. 2007). The use of field data to derive and validate standards has been widely undertaken in freshwaters using a range of statistical techniques (e.g., Crane et al. 2007; Linton et al. 2007). These techniques are readily applicable to soils and may provide further supporting evidence to those who feel that some bioavailability-based approaches do not deliver required levels of environmental protection. Pressure for such confirmatory work is likely to increase as many jurisdictions require cost benefit analysis to be conducted for most proposed environmental regulation and legislation. This work should also force science to be increasingly evidence based and to move away from using default assumptions and uncertainty factors.

There is no doubt that considerable effort could be expended in reducing the number of assumptions (such as the various uncertainty factors) that are used. An equal effort could be expended to generate data that would permit all, or the vast majority of, SQSs for TEs to be derived in a similar manner. Consideration should be given as to which TEs should be given priority as addressing either of the above would require considerable resources.

In conclusion, there is considerable scope available for international harmonization of the data, data quality assessment methods, SQS derivation methodologies, and SQS implementation. However, for this to occur, goodwill is required by all participating parties, and jurisdictions must be willing to cooperate or preferably fund an international organization that has been given the task of harmonization, data collection, and data quality assessment. Such international harmonization will not mean that a single set of SQSs will be derived for the entire Earth. The goal would be to harmonize as far as possible the scientific aspects, leaving individual jurisdictions the flexibility to modify the basic methodology to suit their regulatory framework, local environmental conditions, and/or policy decisions.

REFERENCES

Carlon C. 2007. Derivation methods of soil screening values in Europe: a review and evaluation of national procedures towards harmonisation. Report EU 22805-EN. Ispra (Italy): European Commission, Joint Research Centre. 306 p.

Crane M, Kwok K, Wells C, Whitehouse P, Lui G. 2007. Use of field data to support European Water Framework Directive Quality Standards for trace metals. Environ Sci Technol 41:5014–5021.

De Jong F, Verbruggen E, Vos J. 2007. The use of field studies in derivation of environmental quality standards. Poster at SETAC Europe 17th Meeting, Multiple Stressors for the Environment and Human Health Present and Future Challenges and Perspectives. 20–24 May. Porto (Portugal).

Environment Agency. 2009. Using biotic ligand models to help implement environmental quality standards for metals under the Water Framework Directive. Science Report SC080021/SR7b. Bristol (UK): Environment Agency.

Linton TK, Pacheco MA, McIntyre DO, Clement WH, Goodrich-Mahoney J. 2007. Development of bioassessment-based benchmarks for iron. Environ Toxicol Chem 26:1291–1298.

Umweltbundesamt (UBA). 2008. Incorporation of metals bioavailability into regulatory frameworks. Report UBA-FG IV 2.3. Dessau (Germany): UBA.

Abbreviations

ABA	Absolute bioavailability
ABC	Ambient background concentration
ACR	Acute–chronic ratio
AF	Assessment factor
AFo	Oral absorption factor
Al	Aluminum
AM	Arbuscular mycorrhizae
As	Arsenic
As_2O_3	Arsenic trioxide
As_2S_3	Arsenic trisulphide
ATP	Adenosine triphosphate
B	Boron
BaC_{12}	Barium chloride
BAF	Bioaccumulation factor
BLM	Biotic ligand model
BMD	Benchmark dose
BMDL	Benchmark dose limit
BMR	Benchmark response
BTM	Best tracer model
C	Carbon
C/N	Carbon-to-nitrogen ratio
C_a	Anthropogenic concentration or added concentration
Ca	Calcium
CA	Concentration addition
$Ca(NO_3)_2$	Calcium nitrate
$CaCl_2$	Calcium chloride
Caexch	Exchangeable soil calcium
Cb	Gcogenic (natural) background concentration
Cd	Cadmium
CDTA	Trans-1,2-cyclohexanediaminetetraacetic acid
CEC	Cation exchange capacity
CEFF	Effective concentration
Cl	Chloride
CNS	Central nervous system
Co	Cobalt
COPA	Contaminants for priority action
COPC	Contaminants of potential concern
COPD	Chronic obstructive pulmonary disease

Cr	Chromium
CSF	Cancer slope factor
cTEI	Corrected toxicity enhancement index
Cu	Copper
DAF	Dosimetric adjustment factor
DGT	Diffusive gradients in thin films
DOC	Dissolved organic carbon
DTA	Direct toxicity assessment
DTPA	Diethylenetriamine pentaacetic acid
EC	Electrical conductivity
EC10	Concentration causing a 10% effect in the tested organisms
EC20	Concentration causing a 20% effect in the tested organisms
EC50	Concentration causing a 50% effect in the tested organisms
ECB	European Chemicals Bureau
Eco-SSL	Ecological soil screening level
ECx	Concentration causing a x% effect in the tested organisms
ED10	Dose causing a 10% effect in the tested organisms
EDA	Effect directed analysis
EDI	Estimated daily intake
EDTA	Ethylenediamine tetraacetic acid
EIL	Ecological investigation level
EQG	Environmental quality guideline
EQS	Environmental quality standard
e-SOD	Erythrocyte superoxide dismutase
EU	European Union
F	Fluorine
FAME	Fatty acid methyl ester
FAO	Food and Agriculture Organisation
Fe	Iron
FI	Fumigation incubation
FIAM	Free ion activity model
FIR	Food ingestion rate
GC-MS	Gas chromatograph-mass spectrometry
GI	Gastrointestinal tract
GIL	Groundwater investigation level
H	Hydrogen
H_2SO_4	Sulphuric acid
HAc	Acetic acid
HC5	Hazardous concentration to 5% of the species
HCl	Hydrochloric acid
HED	Human equivalent dose
HEDTA	Hydroxyethyl ethylenediamine triacetic acid

HEI	Highly exposed individual
HF	Hydrofluoric acid
Hg	Mercury
HIL	Health investigation level
HNO_3	Nitric acid
IC50	Concentration causing a 50% inhibition in the tested organisms
ICRP	International Commission on Radiological Protection
ID	Index dose
IEUBK	Integrated exposure uptake biokinetic models
IL	Investigation level
IRIS	Integrated risk information system
ISO	International Organization for Standardization
IVG	In vitro gastrointestinal test
K	Potassium
Kd	Soil partition coefficient
LAF	Leaching-ageing factor
LC	Lethal concentration
LC50	Concentration causing a 50% lethality in the tested organisms
LF	Leaching factor
LOAEL	Lowest observed adverse effect level
LOEC	Lowest observed effect concentration
M^{2+}	Free metal ion
MATC	Maximal acceptable toxicant concentration
MeOA	Mechanism of action
MERAG	Metals Environmental Risk Assessment Guidance
Mg	Magnesium
Mgexch	Exchangeable soil magnesium
Mn	Manganese
Mo	Molybdenum
MRM	Maize residue mineralisation
mRNA	Messenger ribonucleic acid
ms-PAF	Potentially affected fraction for multiple substances
Na	Sodium
$Na_2S_2O_4$	Sodium dithionite
$NaNO_3$	Sodium nitrate
NH_4F	Ammonium fluoride
$NH_4H_2PO_4$	Ammonium dihydrogen phosphate
NH_4NO_3	Ammonium nitrate
$(NH_4)_2SO_4$	Ammonium sulphate
Ni	Nickel
$NiSO_4$	Nickel sulphate
NOAEL	No observed adverse effect level

NOAELADJ	No observed adverse effect level adjusted to a weekly average
NOEC	No observed effect concentration
NSI	National Soil Inventory
NTA	Nitrilotriacetic acid
OC	Organic carbon
OECD	Organisation for Economic Co-operation and Development
OM	Organic matter
OSU	Ohio State University
OSU IVG	Ohio State University in vitro gastrointestinal test
PAH	Polycyclic aromatic hydrocarbon
Pb	Lead
PBET	Physiologically based extraction test
PBPK	Physiologically based pharmacokinetic model
PC95	Protective concentration to 95% of the species
PCR	Polymerase chain reaction
PEC	Predicted environmental concentration
PICT	Pollution-induced community tolerance
PLFA	Phospholipid fatty acid
PM	Particulate matter
PM10	Particulate matter with a particle size diameter smaller than 10 μm
PM2.5	Particulate matter with a particle size diameter smaller than 2.5 μm
PNEC	Predicted no effect concentration
PNR	Potential nitrification rate
PO_4	Phosphate
POD	Point of departure
RA	Response addition
RBA	Relative bioavailability
RBALP	Relative bioavailability leaching procedure
rDNA	Ribosomal deoxyribonucleic acid
RfD	Reference dose
RTDI	Residual tolerable daily intake
Se	Selenium
SIP	Stable isotope probing
SIR	Substrate induced respiration
SO_4	Sulphate
SOD	Superoxide dismutase
SQS	Soil quality standard
SS	Soil solution
SSD	Species sensitivity distribution
SV	Soil screening value
tBLM	Terrestrial biotic ligand model
TDI	Tolerable daily intake

TE	Trace element
TF	Transfer factor
Ti	Titanium
TIE	Toxicity identification evaluation
TTF	Trophic transfer function
TTM	Total toxicity of a mixture
TU	Toxic unit
UEF	Urinary excretion factor
UF	Uncertainty factor
UFP	Ultrafine particulates
US-EPA	United States Environmental Protection Agency
WET	Whole effluent toxicity
WFD	Water framework directive
WHO	World Health Organization
WQG	Water quality guideline
Zn	Zinc
ZnO	Zinc oxide
$ZnSO_4$	Zinc sulphate

Index

Page references in **boldface** type refer to tables.

D

E

S

T

U

V

W

Z

Other Titles from the Society of Environmental Toxicology and Chemistry (SETAC)

Ecological Models for Regulatory Risk Assessments of Pesticides: Developing a Strategy for the Future
Thorbek, Forbes, Heimbach, Hommen, Thulke, Van den Brink, Wogram, Grimm, editors
2009

Linking Aquatic Exposure and Effects: Risk Assessment of Pesticides
Brock, Alix, Brown, Capri, Gottesbüren, Heimbach, Lythgo, Schulz, Streloke, editors
2009

Derivation and Use of Environmental Quality and Human Health Standards for Chemical Substances in Water and Soil
Crane, Matthiessen, Maycock, Merrington, Whitehouse, editors
2009

Aquatic Macrophyte Risk Assessment for Pesticides
Maltby, Arnold, Arts, Davies, Heimbach, Pickl, Poulsen, editors
2009

Copper: Environmental Fate, Effects, Transport and Models: Papers from Environmental Toxicology and Chemistry, 1982 to 2008 and Integrated Environmental Assessment and Management, 2005 to 2008
Gorsuch, Arnold, Santore, Smith, Reiley, editors
2009

Veterinary Medicines in the Environment
Crane, Boxall, Barrett
2008

Relevance of Ambient Water Quality Criteria for Ephemeral and Effluent-dependent Watercourses of the Arid Western United States
Gensemer, Meyerhof, Ramage, Curley
2008

Extrapolation Practice for Ecotoxicological Effect Characterization of Chemicals
Solomon, Brock, De Zwart, Dyer, Posthuma, Richards, Sanderson, Sibley, van den Brink, editors
2008

Environmental Life Cycle Costing
Hunkeler, Lichtenvort, Rebitzer, editors
2008

Valuation of Ecological Resources: Integration of Ecology and Socioeconomics in Environmental Decision Making
Stahl, Kapustka, Munns, Bruins, editors
2007

Genomics in Regulatory Ecotoxicology: Applications and Challenges
Ankley, Miracle, Perkins, Daston, editors
2007

Population-Level Ecological Risk Assessment
Barnthouse, Munns, Sorensen, editors
2007

Effects of Water Chemistry on Bioavailability and Toxicity of Waterborne Cadmium, Copper, Nickel, Lead, and Zinc on Freshwater Organisms
Meyer, Clearwater, Doser, Rogaczewski, Hansen
2007

Ecosystem Responses to Mercury Contamination: Indicators of Change
Harris, Krabbenhoft, Mason, Murray, Reash, Saltman, editors
2007

Genomic Approaches for Cross-Species Extrapolation in Toxicology
Benson and Di Giulio, editors
2007

New Improvements in the Aquatic Ecological Risk Assessment of Fungicidal Pesticides and Biocides
Van den Brink, Maltby, Wendt-Rasch, Heimbach, Peeters, editors
2007

Freshwater Bivalve Ecotoxicology
Farris and Van Hassel, editors
2006

Estrogens and Xenoestrogens in the Aquatic Environment: An Integrated Approach for Field Monitoring and Effect Assessment
Vethaak, Schrap, de Voogt, editors
2006

Assessing the Hazard of Metals and Inorganic Metal Substances in Aquatic and Terrestrial Systems
Adams and Chapman, editors
2006

Perchlorate Ecotoxicology
Kendall and Smith, editors
2006

Natural Attenuation of Trace Element Availability in Soils
Hamon, McLaughlin, Stevens, editors
2006

Mercury Cycling in a Wetland-Dominated Ecosystem: A Multidisciplinary Study
O'Driscoll, Rencz, Lean
2005

Atrazine in North American Surface Waters: A Probabilistic Aquatic Ecological Risk Assessment
Giddings, editor
2005

Effects of Pesticides in the Field
Liess, Brown, Dohmen, Duquesne, Hart, Heimbach, Kreuger, Lagadic, Maund, Reinert, Streloke, Tarazona
2005

Human Pharmaceuticals: Assessing the Impacts on Aquatic Ecosystems
Williams, editor
2005

Toxicity of Dietborne Metals to Aquatic Organisms
Meyer, Adams, Brix, Luoma, Stubblefield, Wood, editors
2005

Toxicity Reduction and Toxicity Identification Evaluations for Effluents, Ambient Waters, and Other Aqueous Media
Norberg-King, Ausley, Burton, Goodfellow, Miller, Waller, editors
2005

Use of Sediment Quality Guidelines and Related Tools for the Assessment of Contaminated Sediments
Wenning, Batley, Ingersoll, Moore, editors
2005

SETAC

A Professional Society for Environmental Scientists and Engineers and Related Disciplines Concerned with Environmental Quality

The Society of Environmental Toxicology and Chemistry (SETAC), with offices currently in North America and Europe, is a nonprofit, professional society established to provide a forum for individuals and institutions engaged in the study of environmental problems, management and regulation of natural resources, education, research and development, and manufacturing and distribution.

Specific goals of the society are

- Promote research, education, and training in the environmental sciences.
- Promote the systematic application of all relevant scientific disciplines to the evaluation of chemical hazards.
- Participate in the scientific interpretation of issues concerned with hazard assessment and risk analysis.
- Support the development of ecologically acceptable practices and principles.
- Provide a forum (meetings and publications) for communication among professionals in government, business, academia, and other segments of society involved in the use, protection, and management of our environment.

These goals are pursued through the conduct of numerous activities, which include:

- Hold annual meetings with study and workshop sessions, platform and poster papers, and achievement and merit awards.
- Sponsor a monthly scientific journal, a newsletter, and special technical publications.
- Provide funds for education and training through the SETAC Scholarship/Fellowship Program.
- Organize and sponsor chapters to provide a forum for the presentation of scientific data and for the interchange and study of information about local concerns.
- Provide advice and counsel to technical and nontechnical persons through a number of standing and ad hoc committees.

SETAC membership currently is composed of more than 5000 individuals from government, academia, business, and public-interest groups with technical backgrounds in chemistry, toxicology, biology, ecology, atmospheric sciences, health sciences, earth sciences, and engineering.

If you have training in these or related disciplines and are engaged in the study, use, or management of environmental resources, SETAC can fulfill your professional affiliation needs.

All members receive a newsletter highlighting environmental topics and SETAC activities and reduced fees for the Annual Meeting and SETAC special publications.

All members except Students and Senior Active Members receive monthly issues of Environmental Toxicology and Chemistry (ET&C) and Integrated Environmental Assessment and Management (IEAM), peer-reviewed journals of the Society. Student and Senior Active Members may subscribe to the journal. Members may hold office and, with the Emeritus Members, constitute the voting membership.

If you desire further information, contact the appropriate SETAC Office.

1010 North 12th Avenue
Pensacola, Florida 32501-3367 USA
T 850 469 1500 F 850 469 9778
E setac@setac.org

Avenue de la Toison d'Or 67
B-1060 Brussels, Belgium
T 32 2 772 72 81 F 32 2 770 53 86
E setac@setaceu.org

www.setac.org
Environmental Quality Through Science®

Milton Keynes UK
Ingram Content Group UK Ltd.
UKHW040054071024
449327UK00019B/553